Niagara through
the Lens
the shots that set
The Standard

FIRE SERVICES IN NIAGARA
1937-1950

from

THE STANDARD COLLECTION
ST. CATHARINES MUSEUM

To those who served in protecting our communities

Niagara through
the Lens
the shots that set
The Standard

FIRE SERVICES IN NIAGARA

1937-1950

from

THE STANDARD COLLECTION
ST. CATHARINES MUSEUM

Researched and Compiled by

BRIAN PHAIR

Looking Back
Press

Vanwell Publishing acknowledges the financial support of the Government of Canada through the Book Publishing Development Program for our publishing activities.

Design: Linda Moroz-Irvine

Published by:
Vanwell Publishing Limited
P.O. Box 2131, 1 Northrup Crescent
St. Catharines, Ontario L2R 7S2
905 937-3100 ext. 829
905 937-1760 Fax
sales@vanwell.com

Published in co-operation with:
St. Catharines Museum
P.O. Box 3012, 1932 Welland Canals Parkway
St. Catharines, Ontario L2R 7C2
905 984-8880
905 984-6910 Fax
museuminfo@stcatharines.ca

Customer Service and Orders:
1-800-661-6136

Library and Archives Canada Cataloguing in Publication
Phair, Brian Spencer, 1987-
    Fire services in Niagara : Niagara through the lens /
    Brian Spencer Phair.

Includes index.
ISBN 978-1-55068-923-5

1. Fire extinction--Ontario--Niagara (Regional municipality)--
History--Pictorial works. 2. Niagara (Ont. : Regional municipality).
Fire dept.--History--Pictorial works. I. Title.

TH9507.N52P53 2008   363.37'80971338   C2008-904857-1

 THIS PUBLICATION HAS BEEN MADE POSSIBLE THROUGH
THE GENEROUS SUPPORT
OF
The St. Catharines Professional Firefighters Association, IAFF Local 485

FRONT COVER PHOTO: Fire at MacQuillen's Drug Store – *St. Catharines Standard*, August 28 / 29, 1949, p. 1. StCM – *The Standard Collection,* S1949.29.3.8. Article page 86.

BACK COVER PHOTOS: *St. Catharines Standard*, February 17, 1938, p.14. StCM – *The Standard Collection*, S1938.11.13.1. Article page 8. • *St. Catharines Standard*, April 14, 1938, p.9. StCM – *The Standard Collection*, S1937.28.2.3. Article page 10. • *St. Catharines Standard*, October 4, 1947, p.3. StCM – *The Standard Collection*, S1947.29.3.1. Article page 62.

DEDICATION: PHOTO, page [ii]: Firefighter Escutcheon from front façade of the Lake Street Fire Hall, July 13, 2008. StCM - T2008.9. Photo by Cameron Phair.

TABLE OF CONTENTS

PREFACE

THE ST. CATHARINES MUSEUM IS PLEASED TO PRESENT the first in a series of books prepared from 65 years of images in *The Standard Collection*.

Serious historians and others with a general interest in our past can be thankful for *The Standard*'s foresight in preserving this collection, and in now making it available for the public's enjoyment and education. This photo archives is of unparalleled importance to Niagara and was donated in 2007 to the St. Catharines Museum. Those at *The Standard / Osprey Media* who made this possible include: Paul McQuaig, Publisher; Julia Kamula, Past Publisher; Andrea Kriluck, Managing Editor; and Blair MacKenzie, Vice-President and Legal Counsel; as well as Denis Squires in the City Solicitor's Office of the City of St. Catharines. We are also grateful to Debi Wiebe, Director of Community Relations for *The Standard*, for all of the good work she does in raising awareness, raising funds, and partnering in the community.

This Collection is the most significant pictorial record of life in St. Catharines-Niagara as seen through the lenses of *Standard* photographers. Every day since 1936 they have been documenting 'news,' but, of equal importance, it was another day of building 'history.' Nineteen photographers have helped to build this community archives. We thank them for their dedication and hard work – it is through books such as this that one can truly be grateful for their commitment in recording 'moments' in time.

The St. Catharines Museum is also very fortunate to have volunteers who share a passion for our past. They donate their time and talents to many worthwhile projects, in this case assembling an interesting look-back at an invaluable and sometimes dangerous service to our communities – our firefighters. To Brian Phair, we extend the Museum's and the community's gratitude for a job well done. Furthermore, this book will also be providing an important source of revenue so that all may appreciate *The Standard Collection* into the future. Proceeds from this publication, and future ones, will be dedicated to the proper preservation and cataloguing of the fragile negatives.

The City of St. Catharines has demonstrated a great commitment to the future of this collection through the hiring of a Cataloguer to document the 600,000 images. As a public museum, we subscribe to professional standards to provide access to and long-term preservation of the collection. This staff position is important to the overall success of what is anticipated to be a 25-year project.

To learn more about our past, and photojournalism at *The Standard*, we encourage you to read: "*NIAGARA THROUGH THE LENS: the shots that set* The Standard." This book is produced from the special exhibition of award-winning photographs and other important images showcased at the Museum from September 13 through November 2, 2008. It also provides insightful summaries of the fascinating careers of *The Standard*'s photojournalists.

And lastly, I would like to thank the Cahills – father Lou and son Denis. Lou started at the paper as a freelance writer in the 1930s, and during the War was employed as one of their senior reporters. In 1982, he began the process of the acquisition of *The Standard Collection* by 'opening doors', a skill which Lou possesses in abundance. His son Denis was hired by *The Standard* in 1965 and he has carried on and shared this support in the ensuing years. Denis became Chief Photographer in 1987, and continues to document life in Niagara for *The Standard.*

Arden Phair
Curator of Collections
July 13, 2008

ACKNOWLEDGEMENTS

THERE ARE MANY PEOPLE BEHIND THIS BOOK, several of whom I have never met. I would like to thank all those involved from *The St. Catharines Standard*, from the photographers of the 1930s, 40s and 50s (whose names are unknown to me), to everyone involved in the donation of the extensive negatives collection to the St. Catharines Museum. As for the St. Catharines Museum, I am grateful for access to *The Standard Collection* and the Museum's many resources, and in particular would like to thank all of the unsung volunteers who are dedicated to organizing and cataloguing such a massive collection. Furthermore, I must mention the work of those at the Mayholme Foundation; without them, my research would have been much more challenging. The James A. Gibson Library of Brock University deserves special mention as I spent countless hours using their microfilm collection of *The Standard* and their microfilm reader/scanners. Moreover, this book would not exist without Vanwell Publishing and their *Looking Back* series; thank you. The editorial input of John Burtniak and the image editing by Mike Conley are also acknowledged in providing "polish" to the finished product. And, a special thank you is also due my brother, Graham, for his research assistance while I was out of town. And, finally, I am very grateful to my father who approached me with the idea for this project and for his constant guidance. *Thanks, Dad*.

Brian Phair
June 11, 2008

THE STANDARD COLLECTION

PROJECTS OF THIS NATURE DO NOT JUST HAPPEN. They are accomplished by means of many supporters, some of whom are recognized in the Acknowledgements. A collection of this scope also requires significant funds for conserving, storing and cataloguing the rare photographic images. In addition to thanking the St. Catharines Professional Firefighters Association as a principal sponsor of this book, we would also like to acknowledge the following for their role in acquiring and processing *The Standard Collection*.

SUPPORTERS

Fundraising / Community
Relations Committee
Ruth McMullan
Bob Gosselin
Chris Irwin
Heather Junke
Dave Lewis
Isabell McGarry
Sheila Morra
Sheryl Thornton
Joan Yurchuk

Museum Volunteers
Mike Conley
Mike Smith
Alex Ormston
Margaret Ferguson
Brenda Zadoroznij
Leonard LePage
Brian Thorne
John Vangameren
Mervyn Cripps
Gail Collins
Donna Ford
Arvind Bhide
Grant Jones
Paul Lewis
Irene Romagnoli
Hugh Learmonth
Brian Phair

Graham Phair
John Burtniak
Blair Walker
Cameron Phair
Milton Conrad
Linda McGarrity
Derek Drake
John Fisher
Denis Cahill
Mike Hamilton

Organizations
HRSDC – Canada Summer
  Jobs (2008)
The Historical Society of
  St. Catharines
Mayholme Foundation –
  Corlene Taylor,
  Brenda Young
Peterborough Centennial
  Museum – Sandra Neale,
  Nicole Christie
St. Catharines Public Library
  – Special Collections
Brock University – James A.
  Gibson Library, Special
  Collections
Cascade Design
The Hamilton Spectator

Other Assistance
Maggie Buckley
Alice Steele
Alicia Stock
Karen Cockerham
Al Teather
Murray Thomson
Mike DiBattista
Andrea Kriluck
Steven Remus
Charles Gervais
Jim Phelps
John Kennedy

Supporters
Mary Elizabeth & Bill Fulton
Marynissen Estates
Creekside Estate Winery
Bloomin' Busy Flower Shop
Frosted Memories
  Cake Designs
Wes Turner
Paul Pattison
George Darte
  Funeral Chapel

GUIDE TO ABBREVIATIONS AND OTHER NOTES

A.R.P.: "Air Raid Precautions." An organization originally established in the United Kingdom in 1924. During the Second World War, they offered protection to citizens from the dangers of air raids, as well as providing other emergency services.

No. 9 E.F.T.S.: "Elementary Flying Training School." No. 9 School was located at the present-day Niagara District Airport and was part of the larger British Commonwealth Air Training Plan (BCATP) which trained pilots, navigators, gunners, and others for air service overseas during the Second World War.

Date Listings: where there are multiple dates in the credit information (e.g. "May 11 / 12, 1937"), the first is the day that the photo was taken and the second is the date that the image appeared in *The St. Catharines Standard*.

A single date indicates the day the photo appeared in the newspaper, though in some instances, it may also be the day that the photo was taken.

Photographers: the individual photographers of most of the images featured in this book are not known. For the most part, the only credit which accompanies the newspaper caption is "Staff Photo, Standard Engraving." The likely photographers are either Jack Williams or Don Sinclair, who were staff photographers at *The Standard* from 1936-41 and 1938-69, respectively.

Titles: Titles are in *Italics* if a title appeared with the photo in the paper. Edits and additions are shown in [square brackets].

Captions: Photo captions are recorded as originally published in the newspaper. Some of the terms in use reflect the time period.

Photo Warning: Some photo reproductions and articles may be disturbing to some individuals.

PHOTO ORDERS

THE CATALOGUING of *The Standard Collection* is a long-term project. Photos will gradually become available for public use after a phased-in program of cataloguing. The entire process will take approximately 25 years.

Images in this book have not undergone conservation treatment nor major digital enhancement. Reprints, as they appear here, are available for purchase by contacting the Museum and referring to the catalogue numbers as shown, e.g. S1937.29.1.1

The Standard Collection does not include negatives for all images which appeared in the newspaper. A non-refundable Research Fee is payable for searches of images which do not have a catalogue reference number. Those wishing to place a photo order should have the following information; the publication date and page number, a photocopy of the image and caption as they appeared in the newspaper, and a Research Fee in payment.

PRESERVATION TRUST

DONATIONS ARE GRATEFULLY RECEIVED to financially support the preservation of this important community archives. Funds received will be deposited into a specially designated account for the preservation, conservation, cataloguing, and promotion of *The Standard Collection*. Please contact the Museum for further details. Receipts for Income Tax purposes will be issued on donations.

INTRODUCTION

THE OPPORTUNITY TO PREPARE THIS BOOK was given to me during a semester off from college. However, upon first accepting the project I had no idea of how long it would actually take, and now as I return to complete my last semester, in addition to schooling, I find myself also trying to wrap up this project.

In the end, I deserve little credit for this book, as most of it has been prepared for me. Let me explain. All of the photos on the following pages are part of *The St. Catharines Standard* negatives collection which was generously donated to the St. Catharines Museum for preservation and cataloguing. It is in the basement of the Museum where this book began, sorting through hundreds of fire-related negatives trying to pick the photos which would make the cut. The next stage consisted of heading to the James A. Gibson Library at Brock University to use their collection of *The St. Catharines Standard*. This was the most tedious part of researching this book as exact dates were not available for several photos. Sometimes I had to look through an entire year of newspapers to find only one photo (and it often seemed like the photo would be the last one at the end of a year). Thanks to the microfilm readers at Brock's Library I could scan the page bearing a photo of interest and then retype the caption explaining the image and share it with you.

I didn't use a hard set of rules for what would make the book or what would not. Obviously large fires would usually make it because they just made better photos, however, other photos have been included for the story behind them. In going through the old newspapers I found it very interesting to discover how much has changed in the Fire Departments from yesteryear to now, while realizing how much is still the same. While equipment has changed dramatically, some practices remain very much the same (e.g. "Fire Prevention Week" in schools).

I hope that while you read this book you learn something new or it simply heightens your admiration for those old "smoke eaters."

Brian Phair
June 11, 2008

St. Catharines Standard, May 11 / 12, 1937, p. 11.
StCM – The Standard Collection, S1937.29.1.1

Thorold Firemen Parade as Colour Guard for Presentation

"Smart in their full uniforms, Protection Hose Company, Thorold, paraded last night as a color guard in the ceremony attendant upon the presentation and consecration of the flags given by Sir John G. Thorold to the town, and deposited in St. John's Church. The brigade, with Chief Ekins in the foreground, and the color bearers, Reeve F. J. Smith, and Deputy-Reeve G. T. Richings, with Mayor W. A. Hutt, in the middle, is shown moving along Claremont Street toward the Church."

St. Catharines Standard, May 12 / 13, 1937, p. 2.
StCM – The Standard Collection, S1937.36.2.2

Prelude to Coronation Parade

"Spectators who gathered near Ontario and St. Paul Streets to watch the coronation parade received an added thrill when a few minutes before the parade was scheduled fire broke out in a warehouse building at the rear of Calderone's fruit store. The firemen, who a few minutes later made a fine appearance in the parade, battled the stubborn blaze. The inside of the building was damaged by the fire. An action shot of the firemen is seen above."

[The Coronation of King George VI and Queen Elizabeth took place on May 12, 1937]

St. Catharines Standard, October 24 / 25, 1937, p. 3.
StCM – *The Standard Collection*, S1937.36.7.1

St. Catharines Standard, October 24 / 25, 1937, p. 3.
StCM – *The Standard Collection*, S1937.36.7.2

Old Virgil Hall Is Levelled By Flames

"Fire yesterday afternoon destroyed the Virgil Hall built about 46 years ago. The fire, believed to have started from the stove, swept through the building and threatened other buildings nearby. Pictures above show the fire in an early stage with flames breaking through the roof and the ruins in a later stage of the fire."

St. Catharines Standard, December 8, 1937, p. 31.
StCM – The Standard Collection, S1937.76.6.1

St. Catharines Firemen Do Their Part in Making it a Very Merry Christmas

"A View of some of the toys which were repaired and repainted by the city firemen in their spare time during the past few months. The toys were collected by the Boy Scouts and brought to the Central Fire Hall, where the workshop was utilized in repair and paint jobs. This year there was no public appeal for toys, the Scouts making their request through the schools. The toys will be distributed to children of needy families."

St. Catharines Standard, December 14 / 15, 1937, p. 1. StCM – The Standard Collection, S1937.36.3.1

City Firemen Had a Smoky Workout Yesterday Afternoon

"A View of the blaze at 33 St. Paul Street West which yesterday afternoon resulted in damage conservatively estimated at $300. The fire was between the ceiling and the floor of the second story and proved rather difficult to fight."

[33 St. Paul Street West – home of David K. (mariner) and Agnes McShannon]

St. Catharines Standard, December 16, 1937, p. 21. StCM – The Standard Collection, S1937.92.5.1

Onwards and Upwards for Thorold Fire Department

"Above is a new truck recently delivered to the Thorold Fire Department at a cost of about $8,500. The truck has already made several fast runs, and tests made subsequently showed it to be entirely satisfactory. At the same time as this rejuvenation program was launched by the town for the equipment, council appointed the first full-time man of what may blossom forth soon to be a full-time fire department of a few men at least. The man to receive the appointment was Bob Gillies, seen above in the truck of which he is the driver."

St. Catharines Standard, January 10 / 11, 1938, p. 1. StCM – *The Standard Collection*, S1938.22.2.1

ACADEMY GUTTED BY FIRE; TO BE REBUILT
Fire Causes $300,000 Loss at Loretto Academy

"Fire which swept through a large section of the Loretto Academy, Niagara Falls, Ont., last night, caused a loss which, it is expected, will reach $300,000. The north wing, newest part of the large stone structure, was completely gutted and the roof was burned from most of the other part of the academy. The scene above was taken when the fire was at its height."

St. Catharines Standard, February 17, 1938, p. 14. StCM – The Standard Collection, S1938.11.13.1

Fire Hall in Thorold 60 Years Old

"The Thorold Fire Hall, built 60 years ago, is reported as badly in need of repair. The recent municipal investigation showed the walls of the hall to be bulging and the flooring beginning to give way, seriously endangering the two fire trucks."

St. Catharines Standard, April 8, 1938, p. 1. StCM – *The Standard Collection*, S1938.23.3.1

Historic DeCew Home Destroyed by Fire

"The historic DeCew home, to which Laura Secord walked from Queenston in the War of 1812, was earlier this morning destroyed by fire. The walls, all that remains of the big stone building, are seen in this picture with some of the few effects of the British family, present occupants, loaded on a wagon in the foreground."

St. Catharines Standard, April 14, 1938, p. 9. StCM – The Standard Collection, S1937.28.2.3

Fire Engine Pumper Now Retired

"Citizens in general will recall the fine service given by the old pumper, shown as it was drawn through the city on the occasion of the Coronation Day celebration last May [12th]. The pumper on more than one occasion averted a disastrous conflagration."

[*This 1905 Steam Fire Pumper is on display at the St. Catharines Museum*]

St. Catharines Standard, April 22, 1938, p. 14. StCM – The Standard Collection, S1938.52.3.1

Beamsville Has Fine New Fire Truck

"This efficient piece of fire fighting equipment was delivered this week to the Beamsville Fire Department. The new truck gives the Beamsville department good pumping equipment which under severe tests surpassed specifications."

St. Catharines Standard, August 20, 1938, p. 1. StCM – The Standard Collection, S1938.23.1.2

Fire Sweeps Through Old Factory

"Fire Fighters had a tough battle late this morning when flames swept through an old frame and tin factory at Welland Avenue and John Street. The building, owned by Canadian Canners, was used as a storehouse. Loss is estimated at $8,000 for the building. At noon there was no estimate of the loss to contents which consisted of packing boxes, 100,000,000 labels, and some canning machinery."

St. Catharines Standard, August 20, 1938, p. 9. StCM – The Standard Collection, S1938.23.1.1

Factory Fire Threatens Section of the City

"Fire of an undetermined cause today demolished a frame Canadian Canners' warehouse at Welland Avenue and John Street, threatened for a time to spread to other buildings in the vicinity. A general alarm was sounded in St. Catharines and Merritton fire department with a truck and 20 men lent aid in preventing spread from the high wind."

St. Catharines Standard, November 2 / 3, 1938, p. 22. StCM – The Standard Collection, S1938.55.1.2

City Firemen Receive Instruction in Gas Masks

"Instruction in the use of gas masks is now part of the regular routine for members of the city fire department. Equipped with eight all-service masks suitable for all types of gas and smoke, the firemen once a week spend a short period in a room filled with sulphur. Pictures above show part of yesterday's gas mask drill…. Three masks were used in sulphur-dioxide gas on an alarm only this week."

St. Catharines Standard, July 18, 1939, p. 1.
StCM – *The Standard Collection,* S1939.20.5.14

Fire Destroys Large Fonthill Nursery Barn

"A hundred tons of hay proved ideal fuel for flames which levelled a large barn on the Fonthill Nursery farm this morning."

St. Catharines Standard, August 17/18, 1939, p. 3.
StCM – *The Standard Collection,* S1939.17.3.1

New Fire Equipment at Fonthill

"A modern piece of fire-fighting equipment was yesterday afternoon delivered to the Fonthill fire department and passed stiff tests with flying colors. The machine, a Bickle-Seagrave engine pumps 420 imperial gallons at 120 pounds pressure and has a triple combination outlet including pump, booster and hose. The new fire truck is shown in this picture with Reeve Nelson at the wheel and Chief Herb Minor standing in the back of the truck."

St. Catharines Standard, September 26, 1939, p. 14.
StCM – *The Standard Collection*, S1939.27.12.1

Thorold Firemen to be Hosts at Annual Provincial Fire Convention

"Thorold Protection Hose Company, No. 1, shown here in their smart black and silver drill attire, are to be hosts next summer to visiting firemen from Ontario at the annual convention of the Provincial Firemen's Association.

Front row, left to right: William Nicol, Alex Riddle, John Howell, Gordon Hay, George Cross, William Fenton, Earl Jacobi, Alex Forsyth, Jack Nicholson, Len Allen, John Farnsworth, Clifford Dell, Robert Thomas, Art Wilkinson;

second row, left to right: James Nicol, Fred Speck, Edward McIntosh, William Henry, William White, Alfred Moss, John Holland, Leslie Sears, John McCreary, Carl Ryckman, Harry Catterall, George Caldwell, William Berry.

In the rear row, Chief Frank Ekins, "Cap," the company mascot, seated beside George Gillis, John Hillman, James Mawdesley."

St. Catharines Standard, November 11 / 13, 1939, p. 8.
StCM – *The Standard Collection*, S1939.20.7.1

Fire Destroys Jordan Station Residence

"Fire Saturday night levelled the residence of Donald Glover, Jordan Station. Lack of water hampered efforts of St. Catharines firemen who responded to a call. In this picture one of the firemen, shielding his face from the heat, is seen playing the lone available hose on a blazing roof."

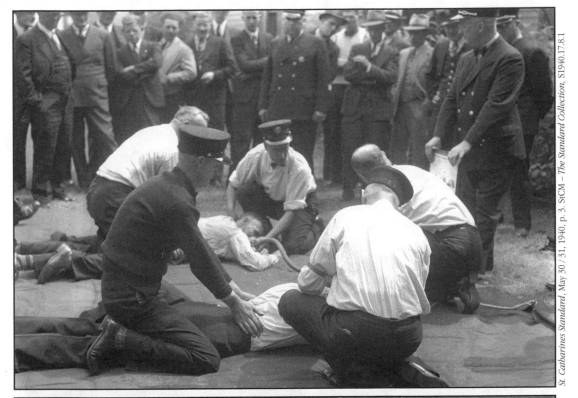

St. Catharines Standard, May 30 / 31, 1940, p. 3. StCM – The Standard Collection, S1940.17.8.1

St. Catharines Standard, May 30 / 31, 1940, p. 3. StCM – The Standard Collection, S1940.17.8.2

Peninsula Firemen Attend Training School Here

"Some 60 members of fire departments and volunteer brigades from all parts of the Niagara Peninsula attended a course in fire fighting and kindred subjects at the St. Catharines Collegiate yesterday morning and then witnessed practical demonstrations by the St. Catharines Fire Department in the afternoon. At top, St. Catharines firemen are showing the proper method of artificial resuscitation using a pullmotor; and lower picture shows the method of carrying an unconscious person from a second storey. The course, under the extension department of Toronto University, the Fire Marshal's department, and the Dominion Association of Fire Chiefs, is continuing today."

St. Catharines Standard, July 20 / 22, 1940, p. 5. StCM – The Standard Collection, S1940.34.6.1

Heavy Loss in Port Colborne Fire

"A Spark from a welding torch which reached an open gasoline tank is believed to have started this fire in Port Colborne at noon Saturday. The large King Street structure, known as the Root building, was practically gutted and the loss is estimated at $25,000. The building was occupied by a transport company, restaurant, shoe repair shop, and apartments. This picture shows the fire at an early stage."

St. Catharines Standard, August 5 / 6, 1940, p. 14. StCM – The Standard Collection, S1940.36.4.1

Firemen's Convention Has Mammoth Parade in Thorold

"Complete with many bands, volunteer fire departments from all parts of the province yesterday marched in the colorful parade which featured their annual convention at Thorold. Although not in competition, the Protection Hose Company, No.1, of Thorold, was the hit of the day in their parade uniforms. Above they are seen just starting a little formation work in the midst of the parade."

St. Catharines Standard, December 4, 1940, p. 23. StCM – *The Standard Collection*, S1940.34.3.1

Fire Department Conducts Auxiliary Santa Claus Shop

"The St. Catharines Fire Department carried on its tradition this year of repairing toys for youngsters whom Santa Claus might not have time to visit. Above, at left, Chief William Chestney and a group of his men help to unload a truck which has just collected some fine toys given by citizens to this worthy cause."

St. Catharines Standard, August 28, 1941, p. 16. StCM – The Standard Collection, S1941.20.2.1

Thorold A.R.P. Fire Equipment

"The Auxiliary A.R.P. unit of Protection Hose Company has recently acquired a small fire truck-trailer which can be attached behind a truck or car and rushed to the scene of a fire in case of emergency. The truck, seen above, carries two tanks of 30 gallons of water each with two hundred feet of hose. The water is pumped out by compressed air."

St. Catharines Standard, May 14 / 15, 1942, p. 24. StCM – The Standard Collection, S1942.13.6.1

Large Unoccupied Barn Levelled by Fire; Loss $3,000

"Fire in a metal clad barn building owned by the Hydro-Electric Power Commission on the Merrittville Highway at Gibson Lake, the former Bixby-Beattie farm, was destroyed by fire with a loss of about $3,000 late yesterday. The fire gained such rapid headway that when the Thorold South volunteer fire department arrived it was virtually destroyed, as shown above."

St. Catharines Standard, July 30, 1942, p. 13.
StCM – *The Standard Collection*, S1942.14.13.1

Thorold South Firemen Hosts to Ontario Convention This Week-End

"Above are members of the Thorold South Volunteer Fire Department, who will be hosts to the annual convention of the Ontario Volunteer Firemen's Association over the week end.

In the picture are: W. Wade, 3rd L.; J. Nelson, 2nd Lt.; J. Aikens, Past Chief; T. Wilson, 1st Lt.; T. Wade, Captain; Harry Nelson, Fire Chief; J. Sentance, Assist. Chief; L.T. Bradley, R. C. Wilcox, C.S. Turner, R. Deplanche.

Back row, left to right, G. Brown, D. Brown, J. Gillis, W. Nelson, Police Chief D. Harold, R. Langlois, L. Zuliani, J. Yarnell, W. Fraser, W. Packer and A. Taylor."

Fire Hall to be Convention Centre

"The Thorold South fire hall and police station, shown above, will be the focal point of the convention of the Ontario Volunteer Firemen's Association this weekend."

St. Catharines Standard, July 30, 1942, p. 14. StCM – *The Standard Collection*, S1942.14.17.1

St. Catharines Standard, September 10 / 11, 1942, p. 8.
StCM – The Standard Collection, S1942.15.1.1

Attorney-General Commended A.R.P. Organization in Welland

"Attorney General G. Conant inspected 300 men and women enrolled in the Civilian Defence Committee, A.R.P., at Welland last night and commended them for their fine appearance and efficiency of organization. Left to right: David Finlay, first aid inspector; Mr. Conant, Mayor T. Harry Lewis, Miss Emily Taylor, Red Cross nursing section, and Walter Thornton, chief of the St. John Ambulance Brigade."

[*After his term as Attorney-General, Gordon Conant served as Premier of the Province of Ontario, 1942-43*]

St. Catharines Standard, November 1 / 2, 1942, p. 1.
StCM – The Standard Collection, S1942.18.2.3

$15,000 Fire Wipes Out St. Catharines Flying Club

"The St. Catharines Flying Club, a civilian organization adjacent to No. 9 E.F.T.S., suffered a $15,000 fire which destroyed one hangar and all their five planes late yesterday afternoon…. Origin of the blaze has not been determined."

St. Catharines Standard, November 14 / 16, 1942, p. 3.
StCM – *The Standard Collection*, S1942.18.3.1

St. Catharines Standard, November 14 / 16, 1942, p. 3.
StCM – *The Standard Collection*, S1942.18.3.2

A.R.P. Proves Efficiency at $3,000 Fire at Black Duck Camp

"Using their pumper on their first actual alarm, the Auxiliary Fire Service of St. Catharines A.R.P. aided greatly in fighting the $3,000 fire at the Black Duck tourist camp west of St. Catharines on Saturday. "If we hadn't had their pumper we'd have lost the building," stated Fire Chief L. Arthur Burch today. The little pumper was placed near Ed White's artificial lake which the fire truck could not reach. It pumped water into a cistern, and the fire truck pumped it back out to fight the fire. At [top], auxiliary firemen man their pumper. [Bottom], the smokey [smoky] fire is shown at its height."

St. Catharines Standard, November 17, 1942, p. 1. StCM – The Standard Collection, S1942.18.4.3

ST. CATHARINES STORES SWEPT
Worst Fire in Six Years Causes More Than $150,000 Damage

"The Worst Fire in St. Catharines since the Welland Vale fire in 1936 caused more than $150,000 damage to four large uptown stores this morning. The blaze, which started at 8.30 this morning, was still burning this afternoon. Firemen and A.R.P. workers are shown working in a pall of smoke at Potter & Shaw and Tamblyn drug stores."

St. Catharines Standard, November 17, 1942, p. 3. StCM – The Standard Collection, S1942.18.4.9

St. Catharines Standard, November 17, 1942, p. 3. StCM – The Standard Collection, S1942.18.4.10

Crowds Watch as All Firemen in District Battle Big Uptown Blaze Today

"Several thousand persons watched St. Catharines worst fire in years which almost totally destroyed the valuable stock of Tamblyn's, Potter & Shaw drug stores, J.H. Cummins optometrist store, and H.C. Wallace & Co. dry goods store this morning…. [Top,] A.R.P. auxiliary firemen climb to the second storey; lower,… firemen jam a hoseline into the basement where the fire broke out."

St. Catharines Standard, December 6 / 7, 1942, p. 1.
StCM – The Standard Collection, S1942.18.7.2

Memorial Church Destroyed Sunday in $75,000 Blaze

"Parishioners watch silently as firemen put out the last of the flames which yesterday morning totally destroyed Memorial United Church. Damage was estimated at $75,000, partly covered by insurance. The fire marshal's department is investigating the cause."

St. Catharines Standard, December 6 / 7, 1942, p. 11.
StCM – The Standard Collection, S1942.18.7.4

Third Disastrous Fire Razes Memorial Church

"St. Catharines third bad fire in as many weeks levelled Memorial United Church early yesterday morning. Above, firemen and A.R.P. auxiliary firemen battle the flames. Only the vestibule at the front and the vestry at the back were left standing."

St. Catharines Standard, January 20 / 21, 1943, p. 15.
StCM – The Standard Collection, S1943.17.1.1

New Fire Truck Arrives for St. Catharines

"A new fire truck and pumper arrived here late yesterday from the factory at Woodstock to make the latest addition to St. Catharines' rapidly growing fire department. The pumper has a 700 gallon-a-minute capacity, and was given a stiff three-hour test at the old canal this morning. Fire Chief L. Arthur Burch is seen above watching the gauges as the pumper forces a strong stream of water through its tank. The truck, a gift to the city from the Water Commission, is to be presented officially this afternoon."

St. Catharines Standard, May 9 / 10, 1943, p. 8.
StCM – The Standard Collection, S1943.14.14.7

Victory Parade Was Colorful

"One of the most spectacular parades in the history of St. Catharines marked the Fourth Victory Loan campaign yesterday. Thousands watched as the military and civilian groups took 45 minutes to pass.... Members of the Great War Veterans' Units, the City Council, and the St. Catharines Fire Department, including an impressive array of Civilian Defence equipment, A.R.P. and St. Catharines Fire Department apparatus, together with about 100 representative women war workers, Boy Scouts, Girl Guides, and new Canadian European national groups, formed the civilian section of the parade."

St. Catharines Standard, March 5, 1943, p. 1. StCM – The Standard Collection, S1943.17.3.1

New Aerial Ladder Arrives for St. Catharines

"The dream of the St. Catharines fire department for many years was realized last night when their new aerial ladder was delivered. A preliminary test was made this morning at the armory, and above, the 85 foot ladder is shown at its full extension towering above the roof of the building. The ladder truck also carries 210 feet of other ladders."

St. Catharines Standard, July 19 / 20, 1943, p. 3. StCM – The Standard Collection, S1943.12.6.1

St. Catharines Standard, July 19 / 20, 1943, p. 3. StCM – The Standard Collection, S1943.12.6.2

Niagara Township's New Fire Hall Opened at St. David's

"In an impressive ceremony attended by about 250 township residents and visiting firemen, Reeve Walter H. Sheppard of Niagara Township last night presented the keys of the new Niagara township fire hall to George Jerman, president of the Firemen's Association. Many speakers praised the new hall, the fire equipment, and the efficiency of the new department. The $5,200 brick hall, with a meeting hall on the second storey, is shown at top. In front are the 500-gallon pump tank which carries water to the fire, and beside it is the A.R.P. pumper from Virgil which relays water to the fire truck. Below, Reeve Sheppard presents the keys to the hall committee. Left to right are Fire Chief Joseph Parnall, Reeve Sheppard, Mr. Jerman and Councillor Wilfred Stewart, the other member of the hall committee."

St. Catharines Standard, August 24 / 25, 1943, p. 3.
StCM – *The Standard Collection*, S1943.17.8.5

St. Catharines Standard, August 24 / 25, 1943, p. 3.
The Standard Collection, S1943.17.8.1

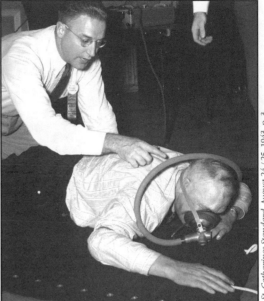

St. Catharines Standard, August 24 / 25, 1943, p. 3.
StCM – *The Standard Collection*, S1943.17.8.2

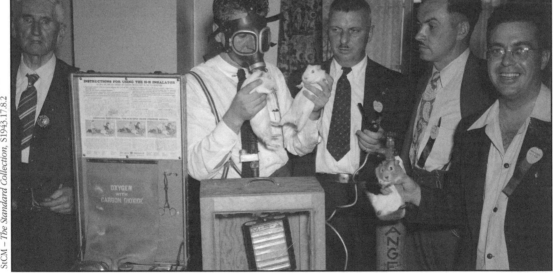

"Busmen's Holiday" for Fire Chiefs' Convention as Fires Are Fought and Lives Are Saved

"The 500 persons attending the 35th annual convention of the Dominion Association of Fire Chiefs here were right at home yesterday as they watched demonstrations of fire fighting and life saving. Manufacturers of equipment offered impressive demonstrations at the Hotel Leonard, convention headquarters, and in a field back of the fire hall....

[Top left], a fire of gasoline and oil, hurling dense black clouds of smoke skyward, is extinguished easily by one man who walked into the blaze behind the protective wall of a fine spray of water."

Top right, Paul Goss, Pittsburgh, demonstrates artificial respiration, using Chief Len Temple, of the emergency squad of the Montreal Light, Heat and Power Co., as patient. The advantages of respiratory equipment are explained.

Lower,... B.W. Catalane, Buffalo, holds aloft two guinea pigs which he had rendered unconscious with dread carbon monoxide gas, only to revive them quickly with an inhalator.

St. Catharines Standard, August 24 / 25, 1943, p. 8. StCM – The Standard Collection, S1943.17.7.1

Fire Chiefs Pay Tribute to Comrades Who Have Passed On

"Following their civic welcome yesterday, delegates to the convention of the Dominion Association of Fire Chiefs marched to Memorial Park where they laid a wreath at the Cenotaph in memory of their members who have passed away. Above, Ernest Wood, York Township, Association President, and L. Arthur Burch, St. Catharines fire chief, are shown laying the wreath."

St. Catharines Standard, September 10 / 11, 1943, p. 3. StCM – The Standard Collection, S1943.17.14.1

Aerial Ladder Truck in Action

"The aerial ladder truck was put in use yesterday, for the second time, when firemen battled a blaze atop the roof of the Monarch Knit Co. The fire was in a frame "pent-house" housing a ventilator fan, and access to the fire was only from the roof. This view shows firemen going up the 45-foot ladder with pump tanks to smother the flames as the girl employees look on from the windows."

St. Catharines Standard, September 25 / 27, 1943, p. 1.
StCM – The Standard Collection, S1943.17.15.5

Victory Bags Salvage Warehouse Swept by $4,500 Fire

"Fire department officials were investigating today the cause of a fire which almost wiped out the salvage warehouse of the St. Catharines Victory Bags Association, North street, late Saturday afternoon. Although the fire spread rapidly through 20 tons of baled and loose paper, it was quickly brought under control when firemen responded to a general alarm. Above, firemen pour water on the paper inside the gutted warehouse, which will be rebuilt within a month."

St. Catharines Standard, September 25 / 27, 1943, p. 3. The Standard Collection, S1943.17.15.4

$4,500 Fire Guts St. Catharines War Salvage Warehouse

"While fire was still smouldering through their burned-out warehouse, directors of the St. Catharines Victory Bags Association met Saturday afternoon and planned to continue uninterrupted collections of salvage from stores and factories. House collections will be resumed in about two weeks, and the warehouse will be rebuilt within a month.... the fire is shown [at] its height, pouring smoke 75 feet skyward.... two... young men... braved the blazing inferno to tie ropes to the four pick-up trucks inside the building so that the trucks could be pulled out. The trucks each suffered about $150 damage. A.C.2 Jack Sharratt of Linwood avenue, said, "The Red Cross has done a lot for me; it's little enough for me to do." Clinton Page, 194 Lake street, saw the smoke of the fire as he was leaving work at the Hayes Steel, Merritton, but arrived in time to help. One of the men helping steer out the trucks burned his hand when he grasped the hot steering wheel of a truck."

St. Catharines Standard, October 16, 1943, p. 1.
StCM – The Standard Collection, S1943.17.16.4

$75,000 Fire at Port Dalhousie Canning Factory

"Thousands of tins of canned goods were destroyed by fire which caused $75,000 damage at the Port Dalhousie canning factory of Canadian Canners, Ltd. at Port Dalhousie early today. Fire broke out at 5.30 o'clock, believed caused from an overheated motor in the elevator. Above, firemen and A.R.P. volunteers are shown pouring water into the steaming ruins, littered with exploded tomato cans."

St. Catharines Standard, October 16, 1943, p. 3.
StCM – The Standard Collection, S1943.17.16.5

Food Destroyed as Canning Factory Burns

"Five-gallon cans of tomato juice, needed for food in wartime, are shown above as they were destroyed in a $75,000 fire at the Port Dalhousie canning factory of Canadian Canners, Ltd., early today. Heavy boilers on the first and second floors, and the weight of a large quantity of canned goods, broke the burning beams and tumbled toward the basement, where ten employees had to scramble out through a cellar window."

St. Catharines Standard, February 27 / 28, 1944, p. 1. StCM – *The Standard Collection,* S1944.16.1.1

Rationed Goods Destroyed in $12,000 Restaurant Fire

"Cases of canned goods, bags of sugar, coffee, tea, butter, and meat were damaged beyond use by fire at the Crystal Restaurant, James St., yesterday afternoon. Total damage, due largely to smoke, was estimated at $12,000. Fire is believed to have started from an overheated furnace. Shown above is the scene at the height of the fire."

*St. Catharines Standard, April 4 / 5, 1944, p. 15.
StCM – The Standard Collection, S1944.16.2.1*

House Fire on Niagara Street Highway

"The half of the double house occupied by Herman Jansen on the farm of W.C. Nickerson, Niagara Street highway, was destroyed by fire shortly before noon yesterday. Mr. Jansen is to be married next week and his home was filled with new furniture for his bride. The furniture was saved, but the Jansen half of the frame dwelling was levelled in the $2,500 blaze. Above, firemen pour water into the ruins of the back half of the double house."

*St. Catharines Standard, September 25 / 26, 1944, p. 1.
StCM – The Standard Collection, S1944.16.4.1*

Fire Razes Landmark in $15,000 Blaze at Wellandport

"Fire starting from defective wiring or spontaneous combustion completely destroyed the chopping mill and feed store in the heart of Wellandport yesterday afternoon. Only excellent work by fire departments from Fonthill and Fenwick checked the blaze which might have wiped out the village. Telephone crews are shown above working this morning to repair communications which were burned out for several hours, isolating the town yesterday."

St. Catharines Standard, September 20, 1944, p. 3.
StCM – *The Standard Collection*, S1944.16.3.1

New Fire Truck for St. Catharines Fire Department

"A new fire truck has been supplied to St. Catharines at a cost to the ratepayers of about $1,700. The new ladder truck is shown above. The city bought a new International engine and chassis, but members of the fire department contributed all the work to make the frame, fenders, running boards, racks, etc., and moved the ladders off the veteran ladder truck to the new machine. The old Reo truck, which has been in service almost 20 years, was worn out, and had not speed enough for the department. The new truck cost $2,100, but $365 was realized when the remains of the old $8,400 Reo were sold. The new truck is heavier, and carried 204 feet of ladders, the usual tools for forceful entry and rescue work, life net, extinguishers, a deck gun for pressure streams, 10 salvage covers, and a moveable spotlight with a half-mile range."

St. Catharines Standard, October 12, 1944, p. 19.
StCM – *The Standard Collection*, S1944.16.5.1

Fire Prevention Taught to Alexandra Students

"Part of the Fire Prevention Week program to familiarize St. Catharines people with fire hazards and how to combat them is being conducted in the schools by the St. Catharines Fire Department. This morning a typical demonstration was given at Alexandra School where more than 700 pupils and teachers evacuated the school in one minute and 57 seconds. Later they were given a lecture on common causes of fire and procedure[s] to follow in case of fire. Above, Fire Prevention Officer N. Batt and Captain Ben Comfort of the Fire Department explain to the pupils how to turn in a fire alarm at a box."

St. Catharines Standard, November 12 / 13, 1944, p. 1.
StCM – *The Standard Collection*, S1944.16.6.1

Exploding Fuel Burns Man to Death in Trailer

"Apparently trying to use gasoline or some volatile fuel to start the fire in his trailer, Wilbert Graham, 34, whose trailer was behind 6 Beecher St., was victim of a double explosion and fire at 5.27 Sunday morning. His clothing was immediately enveloped in flames, and he was terribly burned as he rushed around outside the trailer and finally collapsed as firemen arrived. He died at the St. Catharines General Hospital four hours later, retaining consciousness to the last. Two cans, believed to have held coal oil or gasoline, were found blown apart in the gutted trailer. Fire Chief L. Arthur Burch is shown above inspecting one of the cans outside the trailer."

St. Catharines Standard, January 5 / 6, 1945, p. 3. StCM – The Standard Collection, S1945.18.1.8

Fire at Nu-Bone Corset Company Factory

"Clouds of smoke are shown above pouring from the Nu-Bone Corset Co. factory at the corner of Main and Brock Streets, Port Dalhousie, at the height of the costly fire yesterday afternoon which caused several thousand dollars damage to the structure. Port Dalhousie and St. Catharines firemen battled the flames for two hours before bringing the fire under control, and worked well into the night to extinguish the flames. The top floor was gutted from the blaze, and part of the brick wall on the west side, on the left in the picture, collapsed."

St. Catharines Standard, February 17 / 19, 1945, p. 12. StCM – The Standard Collection, S1945.18.2.2

St. Catharines Man Died of Burns

"Burns suffered Saturday night in a fire at his home, 8 Capner St., caused the death of Ernest W. Davis, former St. Paul St. newsstand proprietor. Fire was confined to the front room and above. Fire Chief L.A. Burch, left, and Deputy Chief Wm. Chestney are shown examining the debris."

St. Catharines Standard, March 5 / 6, 1945, p. 3. StCM – The Standard Collection, S1945.18.3.1

Air Raid Siren Removed from St. Catharines

" "Wailin' Winnie" left her perch above the St. Catharines Municipal Building yesterday afternoon, an indication of the trend of world events. Apparently convinced that this area will not be subjected to air raids, the War Assets Corporation has sold the siren to Louth Township as a fire alarm for $330. The eery sound of the wartime siren sent the shudders through St. Catharines only eight times: once for its initial test; six times for the three blackouts, and once on the occasion of Italy's capitulation. It is to be installed on the roof of the Jordan winery some time this summer. Above, the siren is shown being lowered from the roof as the aerial ladder truck is used as a hoist."

St. Catharines Standard, March 6, 1945, p. 15.
StCM – *The Standard Collection*, S1945.18.4.1

Fire Department Looks to its Past

"An effort to trace the history of fire fighting in St. Catharines is being started by Chief L. Arthur Burch of the St. Catharines Fire Department. In conjunction with his historical research he is collecting souvenirs of the old days. A number have already been presented to the department by Joseph Immel, and Mrs. Fred Gourley of Buffalo, daughter of the late John King, who was one of the oldtimers on the volunteer fire companies. Chief Burch is anxious to receive old records, information, or old pieces of fire fighting equipment from anyone who may be able to help in his search. Above, inspecting four old trumpets and other souvenirs are Fire Captain Ben Comfort, left, and Chief Burch."

[*Some of these artifacts are currently on display at the St. Catharines Museum*]

St. Catharines Standard, October 10 / 11, 1945, p. 13.
StCM – *The Standard Collection*, S1945.18.8.2

English Electric Puts on Fire Fighting Demonstration

"As part of Fire Prevention Week, company fire departments in industries throughout the city are putting on an extra spurt to brush up on their work and interest others in their plant in preventing fires. Through the fine work of plant fire departments co-operating with the city fire department, there has not been a serious industrial fire in the city for many years. Yesterday, the plant brigade of English Electric Co. put on a realistic demonstration of fighting fire with their special equipment, and above, the company smoke eaters are shown wading into a burning oil soaked shack behind a fog nozzle."

St. Catharines Standard, September 22 / 24, 1945, p. 5. StCM – The Standard Collection, S1945.18.6.2

St. Catharines Standard, September 22 / 24, 1945, p. 5. StCM – The Standard Collection, S1945.18.6.3

Farmerettes Burned Out at Vineland

"After sacrificing their vacations to help harvest fruit crops in this district, 18 farmerettes from all parts of the province saw their summer's rewards go up in smoke as fire destroyed their bunkhouse on the farm of Ira Moyer, Vineland, Saturday noon. They lost all clothing and their pay received the night before. At top, some of the girls poke in the rubble for valuables. Below, Ruth Schwartz, Owen Sound, carries bedding to temporary sleeping quarters as the girls determined to stay on the job."

St. Catharines Standard, October '9 / 10, 1945, p. 3.
StCM – *The Standard Collection, S1945.18.7.3*

St. Catharines Standard, October 9 / 10, 1945, p. 3.
StCM – *The Standard Collection, S1945.18.7.2*

Fire Prevention Week in St. Catharines

"This is Fire Prevention Week, and the St. Catharines Fire Department is busy educating young people in city schools of the fine points of preventing the ravages of fire. At top, Fire Chief L. Arthur Burch shows youngsters of Victoria School a little house built on a trailer by the firemen. Half of the house, from cellar to roof, shows how rubbish, faulty wiring, old shingles, dirty chimneys, etc., are fire hazards. The other side of the house is a model of cleanliness and safety. As firemen visit the schools they give fire drill, a lecture, and then recharge fire extinguishers. To empty the old contents of the extinguishers, they let the children operate them to see how easy it is. In lower photo, firemen are shown with the little folk as they tipped the extinguishers over on their tops and handled the hose."

St. Catharines Standard, March 19, 1946, p. 3. StCM – *The Standard Collection*, S1946.25.1.3

Old Reo Pumper to be Replaced This Year

"Equipment of the St. Catharines Fire Department is to be completely modernized this year with the replacement of the old Reo pumper by a new fire truck. Provision was made last night in the city's annual budget for expenditure of $13,750 on the new truck, and a by-law was passed immediately to purchase the truck from Bickle-Seagrave. Shown above, at right, is the old Reo, bought in 1927 for $7,000. It had a pumping capacity of 350 gallons a minute. At the left is one of the new fire trucks, identical in appearance to the new one which will be delivered in June. The truck at left pumps 700 gallons a minute, but the one on order is slightly improved to 750 gallons. The old Reo will be retired because of the difficulty in obtaining repair parts to keep it on active strength."

St. Catharines Standard, May 17, 1946, p. 10. StCM – The Standard Collection, S1946.25.3.1

New Bickle-Seagrave Pumper Delivered and Tested

"The fleet of trucks on the St. Catharines Fire Department is at full strength today after the delivery last night of a new pumper from the Bickle-Seagrave Co. The new pumper, which will be stationed at the Lake St. fire hall, cost $13,750, and will replace the 20-year-old Reo. A sister truck to another new pumper added to the department three years ago, the truck was tested out at the Welland Vale plant this morning. Above, the new truck is shown in foreground, taking in water from the old canal to step up the pressure and force a powerful stream from a nozzle mounted on another truck. Guaranteed to pump 750 gallons a minute, the truck under test this morning pumped 875 gallons."

St. Catharines Standard, June 11 / 12, 1946, p. 1. StCM – The Standard Collection, S1946.25.5.1

$50,000 Damage in Spectacular Blaze at Welland Vale Plant

"Laying down a dense smoke screen that hung over most of St. Catharines, fire early last evening destroyed a storehouse of the Welland Vale Manufacturing Co. In the worst fire in this city in four years, the general alarm blaze, shown above at its height, caused damage estimated at $50,000. A large stock of scarce handles for garden implements, made at the plant, was burned."

St. Catharines Standard, June 29 / July 2, 1946, p. 9. StCM – The Standard Collection, S1946.25.6.1

Fruit Platform and Shed Burn at Grimsby Beach

"A spectacular blaze beside the Queen Elizabeth Way at Grimsby Beach Saturday night destroyed the fruit shipping platform and shed owned jointly by the C.N.R. and A.W. Eikmeier & Son. Damage was estimated at $10,000, and fruit growers of the area were alarmed that the loss might hamper shipments from the heart of the fruit belt at the height of the season. Hydro lines were brought down by the fire, and the Beamsville-Grimsby district was in darkness for half an hour until power was re-routed from Hamilton and Thorold. A night flyer was delayed 20 minutes by the two-alarm fire and hose laid across the tracks. The blaze, which attracted thousands of holidaying motorists, is believed caused by a spark from a passing train."

St. Catharines Standard, July 13 / 15, 1946, p. 1. StCM – *The Standard Collection*, S1946.25.7.4

$10,000 Barn Fire Destroys Crops, Livestock

"Raising a plume of smoke that could be seen for 20 miles, a spectacular fire of undetermined origin caused $10,000 damage at the farm of the Jordan Wine Co., formerly owned by ex-Reeve John C. Dressel of Grantham Township, Saturday afternoon. Two big frame barns, two bulls, farm machinery and five stacks of straw and hay were destroyed, and two horses which escaped had to be shot later. Above, the tile silo leans away from the intense heat as Fireman William Bannan of the St. Catharines Fire Department, pours a stream of water on the blazing hay."

St. Catharines Standard, September 27, 1946, p. 1. StCM – *The Standard Collection*, S1946.25.9.2

St. Catharines Standard, September 27, 1946, p. 1. StCM – *The Standard Collection*, S1946.25.9.6

Arson Attempt Made This Morning

"Scenes at this morning's attempt to burn Club Henley. Left is shown the trail of the burned wick leading from a charred packet of matches to a five-gallon tin of gasoline. Ontario Provincial Police Constable Norman Fach is pointing to the place where the fire started. At the rear is Fire Chief Arthur Burch, beside the can of gas that never properly started to burn. The owner of Club Henley, Robert Thompson, is squatted beside another can of gasoline that was burning when he arrived.... Mrs. Wood is standing in front of the window through which she saw the flames. She is pointing to the location of the fires."

St. Catharines Standard, October 8, 1946, p. 9.
StCM – *The Standard Collection,* S1946.25.12.3

School Children Taught Prevention of Fires

"As part of their Fire Prevention Week campaign, the St. Catharines Fire Department is visiting each school in the city to give fire drill, a lecture, and demonstrate the use of fire extinguishers. Above, children of Glen Ridge School are shown as they listened to Chief L. Arthur Burch this morning. He is pointing out a trailer display of household articles which caused fires in St. Catharines homes, and is explaining what substitutes might be used in the interests of safety."

St. Catharines Standard, October 8, 1946, p. 10.
StCM – *The Standard Collection,* S1946.25.12.1

Fire Prevention Week in St. Catharines

"During their visits to St. Catharines schools this week—Fire Prevention Week—members of the St. Catharines Fire Department are inspecting the schools' fire extinguishers. The old charges are dumped and replaced with fresh solutions. Little tots from the schools are shown how to use the extinguishers, and demonstrate to the older pupils just how easy it is. Above, little folk at Glen Ridge School are shown as they shot streams from the school's extinguishers this morning with the help of firemen."

St. Catharines Standard, October 10, 1946, p. 1. StCM – The Standard Collection, S1946.25.14.1

Port Dalhousie Tests New $6,000 Fire Truck

"A new $6,000 pumper for their fire department arrived last evening at Port Dalhousie, and is shown here as it shoots two strong streams of water during tests on Lock St. this morning. Ordered in the spring, the Bickle-Seagrave[s] truck carries 200 gallons of water in a booster tank, and can pump 500 gallons a minute. Port Dalhousie will house it in their new fire hall, and may also retain their old truck for emergencies."

St. Catharines Standard, November 16, 1946, p. 1.
StCM – *The Standard Collection*, S1946.25.15.5

Dog Saved From $5,000 Sea Cadet Fire

"More than $5,000 damage was done ·by fire at the barracks of the St. Catharines Sea Cadet Corps "Renown" below Burgoyne Bridge shortly before midnight last night. Arthur Hipwell, 70, the caretaker, who was sleeping in the building, narrowly escaped in his night attire. He was restrained from running back into the smoke-filled building to rescue his dog, "Tippy," who was led out by Ed Potter. Mr. Hipwell is shown above with his pet."

St. Catharines Standard, November 16, 1946, p. 8.
StCM – *The Standard Collection*, S1946.25.15.6

Fire at Sea Cadet Barracks

" "Flames were shooting 70 feet in the air," said police who discovered the fire at the St. Catharines Sea Cadet barracks shortly before midnight last night. Fast action by firemen brought the blaze under control and saved the major portion of the frame building…. Sea Cadet officers load live ammunition into their car to remove it from the scene; there were 13,000 rounds for practice purposes in the stores…."

Christmas Tree Lights Start Fire on Chestnut Street

"A flash fire from a spark from Christmas tree lights roared through the apartment of George Stewart, 8 Chestnut St., at 7.30 last night, destroying Christmas presents and furniture, and doing damage estimated at $1,500. About 12 persons, occupants of rooms over MacQuillen's drug store, rushed from the smoky apartments into the sleet storm. Above, at top, firemen work in the gutted room. Below, Fire Chief L. A. Burch and Mr. Stewart examine the skeletons of a floor lamp and the Christmas tree."

St. Catharines Standard, March 11/12, 1947, p. 1.
StCM – The Standard Collection, S1947.79.8.4

Thorold Council Talks Insurance as Arena Burns

"Fire believed starting from a gas stove caused about $10,000 damage to Thorold's arena at midnight last night as town council was meeting to discuss more insurance on town property. Blaze was confined to a dressing room in one corner, but beams and partitions along one end were charred. Above, three youngsters look sadly over the damage which came just as the hockey and skating season neared its close."

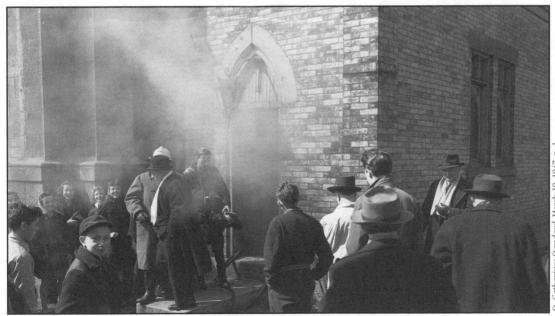

St. Catharines Standard, March 19, 1947, p. 1.
StCM – The Standard Collection, S1947.28.4.1

Smoky Fire Damages St. Catherine's Church at Noon

"Undetermined damage, believed to be slight, was caused by fire in St. Catherine's Church at noon today. The blaze scorched choir vestments and mass candles in a wooden cupboard in the basement at the corner of Church and Lyman Sts. Fire started from hot charcoal in an incense burner hung in the locker after the 10 o'clock mass. Above, smoke pours from the church door as firemen in gas masks combat the blaze."

St. Catharines Standard, March 11 / 12, 1947, p. 8. StCM – *The Standard Collection*, S1947.79.8.2

St. Catharines Standard, March 11 / 12, 1947, p. 8.
StCM – *The Standard Collection*, S1947.79.8.1

Interior of Thorold Arena Damaged in $10,000 Fire

"One of the dressing rooms was gutted and the north end of Thorold's arena was scorched and charred by fire which did $10,000 damage at midnight last night. Fire is thought to have started from [a] gas stove used in the dressing room for heating. At top is shown the heart of the damage in the dressing room, where burned goal pads, gloves and other hockey equipment lie in the ruins. Lower photo shows the exterior of the arena as it appears today. North end (at right) shows scorch marks of searing fire, but metal frame is intact. Arena was opened in 1937 after it was built at [a] cost of $12,000, raised mostly from public subscription."

St. Catharines Standard, April 26, 1947, p. 11. StCM – *The Standard Collection*, S1947.28.5.1

New City Ambulance Ready for Use Today

"A new city ambulance was ready for service today at the central fire hall. The new ambulance, shown above, will comfortably carry four patients at one time, compared with the two older ambulances which each were fitted out for only one patient comfortably. It had been under consideration for several years, due to the occasional accident where three or four persons required hospitalization. Authorized by city council a year ago, the one-ton chassis arrived last fall, and the body was built by St. Catharines Auto Bodies. Members of the fire department put finishing touches to the ambulance this week, and are installing inhalation equipment, a new feature for St. Catharines ambulances."

St. Catharines Standard, May 30, 1947, p. 1.
StCM – The Standard Collection, S1947.28.8.4

Firemen Cheat Robertson School Pupils of Holiday

"Only slight damage was done at noon today by fire which broke out in the tower of the Robertson public school. Firemen, shown above swarming over the building, quickly brought it under control, although smoke had been pouring from the roof when they arrived. School went on this afternoon as usual. Comment from [a] 10-year-old spectator: "Randy Nesbitt turned in the alarm, the old meanie. It would have been fine if it burned down." "

St. Catharines Standard, June 3 / 4, 1947, p. 22.
StCM – The Standard Collection, S1947.80.3.4

Smoke Pall Covers Thorold in $4,000 Mill Fire

"Fire caused by combustion broke out in the Welland Pulp Products mill on Front St. in Thorold last night and caused damage estimated at four thousand dollars. The blaze was confined to the barking room and shaving bins of the mill. The above picture, taken from the west bank of the old Welland Canal, shows the rear of the building with firemen fighting the blaze and huge columns of smoke that swept across the town."

St. Catharines Standard, August 22 / 23, 1947, p. 12. StCM – The Standard Collection, S1947.16.5.2

Fun for Young and Old at Thorold Firemen's Carnival

"The carnival of Protection Hose Co., Thorold, got under way last night at the Regent Street Park with a wide variety of amusements, games, and attractions for young and old. The carnival continues tonight…. Firemen's helmets for the kiddies were a popular feature of the carnival…. Fireman William Henry, Carleton Street, fits hats for two youngsters, whom the photographer learned later were Mr. Henry's daughters, Janice and Sharon."

[*Regent Street Park was later renamed McMillan Park*]

St. Catharines Standard, August 23, 1947, p. 1. StCM – *The Standard Collection*, S1947.28.9.1

August 23, 1947. StCM – *The Standard Collection*, S1947.28.10.1

Fireman Carried Youngster From Home in Midnight Blaze

"William Bannan, member of the St. Catharines Fire Department, was lauded today for his heroic action at 3.30 this morning when he rushed into a blazing home after another fireman and a policeman had been beaten back by intense heat. Fireman Bannan rescued Lynn McClory, 7 [5], from her bed and carried her to safety. Lynn's mother, Mrs. Marjorie McClory, a war widow, leaped to safety from the upstairs triple window shown here. Lynn was in the upstairs bedroom at the left. Damage to the home at 27 South Drive is estimated at $4,000, and cause of the fire has not been determined. The house is owned by Mrs. McClory's parents, Capt. and Mrs. William J. Kirkwood, who were on a lake boat at the time."

Fireman William Bannan.

St. Catharines Standard, September 15 / 16, 1947, p. 1. StCM – The Standard Collection, S1947.28.12.1

Police and Firemen Give $200 to Blind Campaign

"Two $100 bills yesterday were presented by St. Catharines police and fire departments to the Canadian National Institute for the Blind, to purchase radios for the new Niagara Peninsula Home and Centre for the Blind, now nearing completion at the corner of Queenston and Eastchester Sts. The money was proceeds of the so-called "softball" game between the "Flatfoot Specials" and the "Smoke Eaters" last Wednesday night. The two departments are now throwing challenges and counter-challenges at each other in everything from tiddlywinks to hockey. Shown here at the presentation are, left to right: Nat Armstrong, captain of the "Flatfoots"; Inspector Duncan Brown, their manager; L.F. Beattie, chairman of the St. Catharines-Lincoln Advisory Board to the C.N.I.B.; William Bannan, captain of the "Smoke Eaters," and Fire Chief L. Arthur Burch."

Church Burns on "Fuel Sunday"

"Several thousand dollars damage was done [to] the Star of the Sea Church at Port Dalhousie yesterday when fire broke out from an overheated furnace during the 11 o'clock Mass. Worshippers noticed smoke curling up through the floor and registers and filed out quietly. Here, under the direction of Fire Chief Alfred Malton, firemen attack the blaze under the floor inside the Sanctuary. The floor was weakened when basement beams were burned, and services in the church will be suspended until repairs are completed."

St. Catharines Standard, September 28 / 29, 1947, p. 3. StCM – The Standard Collection, S1947.29.2.1

St. Catharines Standard, October 4, 1947, p. 3.
StCM – *The Standard Collection*, S1947.29.3.1

Officers of St. Catharines Fire Department Plan Fire Prevention Program

"Plans for their big annual campaign to prevent loss of life and property through fires are being completed by the St. Catharines fire department, which will conduct Fire Prevention Week next week. Recent changes and promotions in the department have resulted in new officers, and the department's heads are shown here:

seated, left to right, B.F. Comfort, captain; J.R. Harris, deputy chief; L. A. Burch, chief; John Cropper, deputy chief; and Roy Bracken, captain;

back row, Joe Murphy, captain; T.J. Hunt, lieutenant; Harold Moore, mechanic; Harry Oliver, fire prevention officer; John O'Connell, captain; and Charles Garner, lieutenant."

St. Catharines Standard, October 8, 1947, p. 1.
StCM – *The Standard Collection*, S1947.29.5.1

Dutch Veteran Leaps Clear of Blazing Tractor

"In Canada only three months, Martin DeYoung, 42, veteran of the Netherlands army, leaped to safety this morning when his tractor caught fire and overturned on the Middle road, Louth township, near Pay's Hill. Here, firemen extinguish the blaze which almost destroyed the tractor after DeYoung had been unsuccessful in fighting the flames with bags and dirt. He said the tractor caught fire on the road, and as he was turning off the ignition, it veered to the side and overturned into the ditch. He was hauling a trailer loaded with empty grape boxes for T.L. Yungblut, his employer."

St. Catharines Standard, October 6, 1947, p. 9, StCM – The Standard Collection, S1947.29.4.1

St. Catharines' Efficient Fire Department Gives City One of the Lowest Insurance Rates

"Twelve pieces of modern apparatus in the St. Catharines Fire Department, used by a highly trained force of men, have given the city one of the lowest fire insurance rates in the country. The Department this week is conducting its annual fire prevention week to clean up hazards and educate the public to prevent loss of lives and property through fires. Here is shown a line-up of the equipment, which has a replacement value of $110,000. It includes two ladder trucks, one of them an aerial; four pumpers; an A.R.P. pumper bought by the city; a service truck, chief's car, and two ambulances. Not shown is another piece of equipment, the life saving boat and trailer."

St. Catharines Standard, November 4, 1947, p. 1. StCM – *The Standard Collection*, S1947.29.9.1

Child Dies When Fire Sweeps Merritton Home

"Billy Freeman, two-year-old adopted son of Mr. and Mrs. Vern Freeman, 54 Wanda Rd., Merritton, died when fire believed caused from spontaneous combustion swept the top of their one-and-a-half storey home shortly before 8 o'clock this morning. Although flames in the top storey were so intense they drove back firemen, the little tot, sleeping in his crib, died of smoke and fumes. Above, firemen and doctors man an inhalator in an unsuccessful effort to revive him. The child's bedroom was at the front of the house, shown in the background...."

St. Catharines Standard, November 7 / 8, 1947, p. 1. StCM – *The Standard Collection*, S1947.29.11.8

FIREMEN OVERCOME IN $200,000 BLAZE

Spectacular 3¹/₂ Hour Fire Causes $200,000 Damage in Business Block

"Thousands of persons jammed St. Paul and Academy streets last night when a stubborn $200,000 fire resisted firemen from 7 to 11 p.m. Heaviest damage was in the basement of Seigel's shoe store, where Christmas stock had just arrived. Wallace's house furnishings store next door suffered heavy loss from fire, smoke, and water. Here, firemen use gas masks and guide ropes to fight the blaze at its height. Heavy, acrid smoke blanketed the street and overcame four firemen."

St. Catharines Standard, December 22 / 23, 1947, p. 1. StCM – The Standard Collection, S1947.29.13.4

Christmas Gifts Burned in $40,000 Fire at Beatty Store

"Fire of undetermined origin swept rapidly through the basement of the Beatty Sales and Service store at 314 St. Paul St. shortly after noon yesterday. The general alarm fire, second in St. Catharines in six weeks, was brought under control an hour later, with damage estimated at $40,000 to recently redecorated building and valuable stock of large Christmas items. Included in the loss were about 125 radios, some of them big combination record players. One end of the basement was filled with wrapped merchandise already paid for and awaiting delivery as Christmas gifts tomorrow. The manager had planned to give unsold toys to needy children at closing time Christmas Eve, but all that remained of the toy department was a charred shelf of ashes and twisted wire. Here, smoke is shown pouring from the front of the building as firemen, who needed gas masks, pour water in from front and back to drown the blaze."

St. Catharines Standard, January 19, 1948, p. 1. StCM – *The Standard Collection*, S1948.32.2.2

One Overcome in $20,000 Blaze at Avenue Pharmacy

"After giving the alarm for a $20,000 fire which swept the Russell Avenue Pharmacy shortly after 4 o'clock this morning, Wills B. Jamieson, 73, collapsed from the dense smoke in his upstairs bedroom and had to be carried down a ladder by firemen. In hospital today he is recovering from shock and smoke. Mr. and Mrs. Alvin R. Seager, who also occupied an apartment over the drug store, were able to flee next door to safety. Cause of the fire, which started in the basement and burned through the floor into the store, has not been determined. Here, firemen hampered by stiff and frozen hydrants, pour water into the front of the store during their two-hour battle."

$15,000 Fire Guts Grocery Store

"A $15,000 fire believed started from overheated furnace pipes last evening sent clouds of acrid smoke billowing over [the] corner of Queenston and Hartzell [Hartzel] roads. Extensive damage was done [to] the stock of Howard Foxton's suburban corner grocery store and the newly-decorated apartment upstairs. Low water pressure in east St. Catharines hampered efforts of firemen. Here, smoke is shown pouring from the brick building at the height of the blaze."

St. Catharines Standard, January 18 / 19, 1948, p. 9. StCM – *The Standard Collection*, S1948.32.1.1

St. Catharines Standard, January 21, 1948, p. 1.
StCM – *The Standard Collection*, S1948.32.3.15

Two Babies Die as Home Burns When Left Alone

"Two little babies, daughters of Mr. and Mrs. Reuben Flowers, died in a fire which left only this charred skeleton of a house at 39 Garnet St. shortly before noon today. Left alone when their mother went to a cleaners, the little victims died when an overheated coal stove in the kitchen ignited the flimsy house which their father had built with his own hands. Dead are Joyce Elaine, 4 months old, and Marilyn Lorraine, two years. A brother and two sisters were at school at the time."

St. Catharines Standard, January 21, 1948, p. 8.
StCM – *The Standard Collection*, S1948.32.3.16

Mother Returns Home to Find House Burned and Two Babies Killed

"Leaving her two youngest children sleeping while she went shopping this morning, Mrs. Reuben Flowers, 39 Garnet St., returned an hour and a half later to find the home gutted by fire from an over-heated stove, and the two babies burned to death…. Neighbors noticed the flames, were beaten back by heat and smoke as they tried to enter. Below are surviving members of the family: Mr. and Mrs. Reuben Flowers, with Shirley Anne, 8, Melvin, 6, and Joan Marie, 7. Mr. Flowers had built the house himself and the family moved in only a few months ago. They lost a son four years ago. No insurance was carried on house or contents."

St. Catharines Standard, January 26, 1948, p. 3.
StCM – *The Standard Collection*, S1948.32.7.6

Sympathetic Citizens Hold Bee to Help Flowers Family

"Tide of assistance to the family of Reuben "Sunny" Flowers, whose home and two youngest children were burned last Wednesday, continued to roll in today. Financial donations totalled $1,900 from a city-wide response, and as much more from a canvass at McKinnon Industries where Flowers worked. Saturday, an old-fashioned "bee" was held on Garnet St. to clear away debris of the charred home and prepare for the new home guaranteed by the Kiwanis Club supported by the public fund.... volunteers swarm over the charred wreckage as a blizzard raged.... Even as the building was being torn down plans were going forward for construction of the new home.... Ald. Richard Robertson, [is] president of the Kiwanis Club which sponsored the fund; Kiwanian Jim Stork, ... is donating his services as contractor...."

St. Catharines Standard, February 4, 1948, p. 1.
StCM – *The Standard Collection*, S1948.32.8.1

Five Saved When Fire Razes House

"The fact that Bruce Charlton, 15, slept in this morning instead of getting up at his usual time to deliver morning papers possibly saved the lives of five persons in his home. Bruce woke up a little after 5 o'clock, dressed, and started for the back of the house to cover his paper route [when] he saw flames bursting through the lean-to kitchen at the back of the house. Bruce ran back to wake his step-father, Mr. Trefonavitch, and other members of the family, and then raced next door to the home of Reg. Merrick to sound an alarm."

St. Catharines Standard, May 2 / 3, 1948, p. 13.
StCM – *The Standard Collection*, S1948.24.7.1

Firemen At Niagara-On-The-Lake Present Ambulance To Town

"A new ambulance was presented to the town by the Niagara-on-the-Lake Fire Department yesterday. To man the ambulance skilfully, firemen took a six-week course in first aid, and 19 were presented with St. John Ambulance certificates by Clifford Luce, St. Catharines superintendent. Above are firemen who passed the course:

front row, left to right, Fire Chief Don Sherlock, Edward Smith, R. W. Richardson, Frederick E. Ball, F. W. Curtis, C.R. Riches, J.A. Nesbitt, E. Saunders, N.P. Marino, F.E. Garrett;

back row: Alex Russell, Carmen Gould, Walter M. Reid, E. A. Bradley, J.C. Redhead, C.A. Berge, H. Sherlock, A.H. Awde, W. B. Summers, and William T. Bishop."

St. Catharines Standard, May 2 / 3, 1948, p. 13.
StCM – *The Standard Collection*, S1948.24.7.4

"The fire department conducted a financial campaign to raise more than $2,000 from Niagara citizens to buy and equip the new ambulance. It will be used to serve both the town and neighboring area. Since it was delivered in December it has had 36 calls. Fire Chief Don Sherlock in a brief ceremony presented the keys to Mayor Lew McConkey, who is shown here turning them back to the department for their custody."

St. Catharines Standard, August 10/11, 1948, p. 1.
StCM – *The Standard Collection*, S1948.33.2.1

Basket Exchange Provides Ideal Tinder for Furious Blaze

"Fire wiped out the St. Catharines Basket Exchange on Rykert Street, Grantham Township yesterday afternoon. Damage, estimated at $2,000, included loss of frame building, 2,000 baskets and hampers, and a ton and a half of grease dripping which would have been sold today to soap factories. Here, the blaze is shown at its height just before the building collapsed. Firemen stand with dead hoses as the department tried to conserve the meagre supply of water from a well by baling it into their booster pump and squirting it from small hoses only to prevent spread of the blaze from the doomed building. Owner Harold Shapiro, 131 Geneva St., said he was burning broken baskets and thought an ember must have flown onto a pile of flimsy boxes, spreading to the building. It was only partly insured."

St. Catharines Standard, September 27 / 28, 1948, p. 1. StCM – *The Standard* Collection, S1948.33.4.4
Photo by Don Sinclair

$300,000 Fire Guts Heart of Ridgeway Business Section

"Five business places were demolished by fire late yesterday afternoon in the heart of Ridgeway, causing damage estimated at $300,000. The town has no water system, and seven fire departments, including Buffalo, had to rely on wells, railway tenders of water rushed to the scene, and tank trucks of water carted from Crystal Beach. At one time residents were warned to evacuate the town if the wind changed. Here, some of the huge crowd that blocked the main street watches as the fire continues to lick up the ruins of the stores."

St. Catharines Standard, October 4 / 5, 1948, p. 16. StCM – *The Standard Collection*, S1948.33.5.1

Merritton Firemen Demonstrate New Equipment During Campaign

"This is Fire Prevention Week, and across the country fire departments, both full-time and volunteer, are devoting their energies to saving the public the expense and heart-break of fire. Yesterday the Merritton Fire Department demonstrated methods of fighting various types of fires for the high school students. Here, Lieut. Tom Rapson can stand at a safe distance to fight a raging fire of oil and gasoline. He uses a foam generator, a little tank strapped to his back, to spray smothering foam like a snowstorm onto the blazing surface. There are few of these pieces of equipment in the district."

St. Catharines Standard, October 4 / 5, 1948, p. 1.
StCM – The Standard Collection, S1948.33.5.2

Firemen Launch Fire Prevention Campaign Through Schools

"This week marks the anniversary of the great Chicago fire of 1871, when Mrs. O'Leary's cow kicked over a lantern and started a conflagration which raged across the city for four days. Each year the anniversary is marked by firemen who try to impress on the public the importance of cleaning up dangerous rubbish, having furnaces and chimneys in good order, and taking other simple precautions which may save lives and property. Most effective is their campaign in schools, where children are impressed and carry the message home. Here, the St. Catharines fire department is shown during a demonstration yesterday at Connaught School. Children are always keenly interested in fire trucks and equipment, and Deputy Chief R. Harris is showing them a handy collapsible ladder. They were also intrigued with the radio system recently installed in the chief's car."

St. Catharines Standard, October 5 / 6, 1948, p. 10.
StCM – The Standard Collection, S1948.33.6.4

Fire Prevention Week Marred by Fatal Blaze

"Firemen this week have been preaching the danger of fire when not carefully controlled, and Fire Prevention Week in St. Catharines got away to a bad start with a graphic illustration for their campaign. "Bobby" Brown, 68, colored, yesterday was caught in exploding gasoline and died seven hours later after he tried to set fire to 68 Page Street to get revenge on his landlady for serving an eviction notice against him. Two fires, an hour apart, threatened the flimsy frame building. Here, smoke swirls about it during the second blaze. Damage from [the] fire was not great...."

St. Catharines Standard, November 11 / 12, 1948, p. 1. StCM – The Standard Collection, S1948.33.8.8

$100,000 Spray Plant Fire at Beamsville Heavy Blow to Farmers

"Only company in Canada making microfine sulphur sprays and dusting powders, the Bartlett Spray Works, was crippled late yesterday by a $100,000 fire that destroyed the sulphur grinding mill and store-house. The ruins are shown here, with blue sulphur dioxide fumes still curling up from the debris. Fire started in [the] grinding mill at left, where an explosion ripped open a sealed steel drum used for rolling sulphur after it was ground. Fruit and vegetable growers across Canada depended upon the plant for sulphur sprays. Norman Bartlett, the owner, said his plant made the best sulphur in the world due to special machines he had developed for a fine grind."

*St. Catharines Standard, December 13 / 14, 1948, p. 3.
StCM – The Standard Collection, S1948.84.8.1*

Veteran Smoke-Eaters Honored at Merritton Firemen's 60th Birthday

"Guests of honor at a banquet which last night marked the 60th anniversary of the founding of the Merritton fire department were eight old-timers who still race to the scene every time the sirens blow. At a banquet, floor show, and dance at Feor's Rose Room, long service pins were presented to the eight veterans who have been on the department from 20 to 44 years, and are still active members. Here, Wright Partington, member for 44 years, and Mrs. Partington, use the engraved firemen's axe presented to them to cut the anniversary cake. Other veterans are,

front row, left to right, John Riches, Bill Teasdale, Mr. and Mrs. Partington, A.T. Smith, and Jimmy Rennie;

Back row, John Jackson, Jack Rountree, member of the department and secretary of the Ontario Firemen's Association, who presented the pins; Fire Chief L.A. Burch, St. Catharines, who made the presentation of the engraved axe; Chief Art Tuckwell, of Merritton, and Claude Richardson."

St. Catharines Standard, January 20 / 21, 1949, p. 12.
StCM – The Standard Collection, S1949.27.1.1

St. Catharines Standard, January 20 / 21, 1949, p. 12.
StCM – The Standard Collection, S1949.27.1.2

Fire Damage $2,500 at Loblaw Queen Street Store

"Fire starting in packing boxes inside the back door did damage estimated at $2,500 late last night at Loblaw's store on Queen Street. About $1,000 worth of damage was from the dense smoke which packed the store and affected perishable foods. A pedestrian who turned in an alarm from a fire box waited in vain for the trucks to arrive, as a spring had broken in the box; he then ran another block to [the] central fire hall to complete the alarm. Here, at top, firemen enter one of the rear doors as the smoke pours out of the opening. Below, a fireman stamps out blazing boxes tossed out of the building."

St. Catharines Standard, February 21, 1949, p. 1. StCM – The Standard Collection, S1949.27.5.1

Sod Turned for New $125,000 St. Catharines Fire Hall

"Something council has mooted for 25 years—removal of [the] Central Fire Hall from St. Paul St.—was started today when Ald. Harry Robinson, chairman of the Fire Committee, turned the first sod for a new $125,000 fire hall on a triangular block between Geneva and Niagara streets, near St. Paul. It is expected the hall will be in use by fall, and will be one of the most modern in Canada. After Ald. Robinson turned the spade, Mayor Richard Robertson decided he would try something he "always wanted to do," and climbed to the controls of the steam shovel where he went through the motions of starting the excavation. Here, members of council watch as the mayor tinkers with the levers. Others, left to right, are: Ald. Wilfred Bald, Ald. W.W. Walker, Architect R.I. MacBeth, Fire Chief L.A. Burch, Ald. Harry Gale, Ald. T.R. BeGora (behind Ald. James Barley); Ald. John Franklin, City Clerk H.H. Smith and Ald. Robinson, standing on caterpillar."

St. Catharines Standard, April 3 / 5, 1949, p. 3.
StCM – *The Standard Collection,* S1949.28.1.1

Garden City Beach Whips Up Fire Department in One Week

"In little more than a week ratepayers of the Garden City and Municipal Beach area, Port Weller, organized a volunteer fire brigade, and bought and paid for their own truck. Members are shown here as they proudly display the little truck which they bought for a nominal sum from Grimsby after that department got a new truck. They are members of the Niagara District Firemen's Association, and will act as an auxiliary to the St. Catharines department, which has jurisdiction throughout Grantham Township. Merritton firemen are giving them valuable pointers on the organization of a volunteer department, although many of the members have had previous experience with industrial brigades in St. Catharines. The truck carries a 66-gallon chemical tank to serve the 300 homes in their congested beach area, and they plan to add more tanks, ladders, hose, alarm and other equipment. Although built up now, the area has applied for water service, which not only would enlarge the community, but would also add to the facilities of the fire brigade. The truck, which was completely paid for after residents donated to a fund as the truck paraded Sunday, will answer calls to fight fires until the St. Catharines department can make its five-mile run from the city."

Gassed in Car, Tot Recovers

"Anita Jean Pendykosby [Pendykosky], 13 months old daughter of Mr. and Mrs. Ed. Pendykosky, 215 Niagara Falls Road, Thorold South, is shown with Fireman Bob Gillies of the Thorold Fire Department who revived her after she was gassed with carbon monoxide Saturday night. The little girl is now completely recovered from her experience, but despite Gillies' effort to make friends, she showed no gratitude. Out for a ride in the family car Saturday night she became sick, and while her father tried to find a place to park, her mother took her from the car for fresh air. The youngster suddenly went limp in her mother's arms, and Mrs. Pendykosky, ran with her to the fire hall. Fireman Gillies administered oxygen with the inhalator and revived her. "She was all blue when I saw her. I thought she was dead," he said. Others in the car suffered little effect from the carbon monoxide which seeped into the car from a faulty exhaust."

St. Catharines Standard, April 3 / 5, 1949, p. 8.
StCM – *The Standard Collection,* S1949.86.1.1

St. Catharines Standard, April 1 / 2, 1949, p. 5.
StCM – *The Standard Collection*, S1949.85.10.2

Merritton Fire Truck Is Tested

"Merritton's fire fighting facilities took a formidable increase yesterday, when the new $14,000 fire truck ordered approximately a year ago, arrived. The truck was tested in the afternoon at the settling pond behind the Hayes Steel, and according to Fire Chief Art Tuckwell, he was more than satisfied with the test, as the truck performed well. The test was under the charge of Jim Rennie, and the truck was pumping for two hours at 125 pounds pressure, shooting out 620 gallons a minute from two hoses. Although the figures were not available this morning, an even higher test was given [to] the truck. The picture above shows the truck, with one stream of water shooting out of the hose visible at the right. Mayor Skipper and Councillor Ricci are shown at the right discussing the features of the new truck."

St. Catharines Standard, April 6 / 7, 1949, p. 10.
StCM – *The Standard Collection*, S1949.28.2.1

Extensive Damage Is Caused in Fire at Eastchester Home

"Fire early yesterday afternoon, caused $3,500 damage to the furniture and home of Fred Stevens, Eastchester Avenue, adjacent to the Homer dry dock. Fire was caused when spilled oil from an oil heater was ignited by heat from a stove on the ground floor below the heater. Firemen… are seen using pails of water to douse the smouldering remains of a chesterfield suite. Contents which were saved by Mr. Stevens and neighbors are seen at the right…."

St. Catharines Standard, June 9 / 10, 1949, p. 23. StCM – The Standard Collection, S1949.28.4.1

Firemen Battle Spectacular West City Blaze

"A staff photographer caught this photo just as the charred skeleton of what was once a lean-to at the rear of 42 Chetwood Street, crumbled before the streams of two lines of hose. Hordes of school children flocked to the fire yesterday afternoon at 4.30 when dense clouds of heavy black smoke billowed from the rear of the house. Minutes after this photo was taken not even a wisp of smoke spewed from the sodden heap of wreckage which covered the back yard. The rear shed was a total loss but damage to the rest of the house was relatively slight."

St. Catharines Standard, June 19 / 20, 1949, p. 3. StCM – *The Standard Collection*, S1949.28.5.1

Stubborn Blaze Defies Thorold Firemen at Spun Rock Wool Mill

"Protection Hose Company men at Thorold had just returned from the Decoration Day service yesterday when they were forced to don hot helmets and raincoats over their uniforms and struggle 3 hours with a stubborn fire at the Spun Rock Wools plant…. Damage to building and contents was estimated at $10,000."

St. Catharines Standard, June 21, 1949, p. 1. StCM – *The Standard Collection*, S1949.28.6.5

Fire Fighters Open Ontario Convention Here

"Firemen from all parts of Ontario converged on St. Catharines today for the 27th annual convention of the Provincial Federation of Ontario Fire Fighters which opened this morning at the Hotel Leonard. Convention sessions continue until Thursday night, as tours and special events are arranged for the ladies. After a civic welcome this morning delegates paraded behind pipers and fire trucks to the Cenotaph to lay wreaths. Here, Rev. Christopher Loat offers prayers at the brief service. In front are William Bannan, president of the St. Catharines local, and Lorne MacRosti, Ottawa, provincial president. Uniformed members of the St. Catharines department were followed by sweltering shirt-sleeved delegates."

St. Catharines Standard, August 3 / 4, 1949, p. 1. StCM – *The Standard Collection*, S1949.28.9.1
Photo by Don Sinclair

Tank of Gasoline Ignites to Burn N.S. & T. Bus

"A full tank of gasoline—65 gallons—ignited at 7 o'clock last evening as one of the newer N.S. & T. buses was being gassed up in the company yards on Welland Ave. Here, company workmen rush up foam fire fighting equipment as the blaze spreads rapidly from the rear end gas tank through the inflammable upholstery and cushions. Dense cloud[s] of black smoke and orange flame shooting high in the air attracted hundreds of spectators. Although nozzle of gas hose was burned off, fire did not follow hose back to underground storage tank. Firemen said that explosion was prevented by fact that gas tank on bus was full, leaving no room for explosive gasoline vapor. Cause, being investigated today, may have been static electricity, or spark from hose nozzle striking side of gas tank; $8,000 damage was done to $15,000 bus."

St. Catharines Standard, August 17, 1949, p. 1.
StCM – The Standard Collection, S1949.29.1.1

Firemen Halt Dangerous Blaze in Heart of Business Section

"Fire discovered at 5 o'clock this morning in Erwinne's ladies' shop at 159 St. Paul St. threatened for a time to sweep through St. Catharines' main business block, repeating the disastrous $150,000 fire in nearby Wallace's and Potter & Shaw stores eight years ago. All available fire apparatus in the city was rushed to the scene and systematically checked and extinguished the blaze before it could spread. Fire started from faulty wiring and damage, largely from smoke, was estimated at $6,000. Here, the aerial ladder is used to attack the ceiling blaze from above, as wisps of smoke seep from third floor gables."

March 25, 1949.
StCM – The Standard Collection, S1949.27.9.1

Lake Street Fire Hall, Station No. 2

113 Lake St. (cor. Albert St.) • Built 1913 / 1942 (south bay for pumper) / 1976 (rear extension for ambulances) • Architects: Thomas H. Wiley / Lionel A. Hesson / Min. Gov't Services • Designated 1997

The Lake Street Fire Hall was officially opened December 15, 1913 by Mayor Dr. W.H. Merritt. The building cost $9,964.90, plus architect fees of $450. It had the most up-to-date features of the time and included stalls for two horses which would automatically open after an alarm was triggered. Harnesses would automatically drop onto their backs, ready for service with their horse-drawn hose wagon (horses were removed from St. Catharines fire halls in 1925). Station No. 2 then had a 1920 four-cylinder Reo truck. In 1942 a second vehicle bay was added to the south side of the building which accommodated a 1943 Bickle-Seagrave pumper (above left). To the right is a 1944 International ladder truck. The building was lengthened in 1976 to house four ambulances. After Municipal Amalgamation in 1961, the Lake Street Fire Hall was abandoned (May 26), and its equipment transferred to Merritton and Port Dalhousie. Thereafter, the building was used by the Emergency Measures Organization (1962-69), Region Niagara Public Works (1973-75), St. Catharines & District Ambulance Service (1976-96), Down to Earth Pottery (1998-2006), and 2M Architects Inc. (2006 - present).

St. Catharines Standard, August 22, 1949, p. 1. StCM – The Standard Collection, S1949.29.2.1

Volunteers Brave Flames at Firemen's School at Merritton

"About 175 members of volunteer and industrial fire brigades throughout the Niagara Peninsula attended the first annual firemen's training school of the Niagara District Firemen's Association at Merritton Saturday and Sunday. In [the] first school of its kind in the province, firemen get [the] feel of new apparatus and learned [a] wide variety of subjects by practical application. Here, smoke-eaters crouch as they haul hose in to smother [a] gasoline fire."

Volunteer Firemen Enthusiastic Over Chance to Learn Fire-Fighting in Practical School

"Contrasted with firemen's training schools of the Ontario Fire Marshal's Department, where only lectures are given, the school attended by 175 Niagara District firemen at Merritton Saturday and Sunday gave the volunteers practical work. They learned a wide variety of subjects by actually doing them in 10 different groups which rotated.... They learned that three of the masks are good only for smoke, while… one… is an all-purpose mask developed during the war....William Hagerty, of the Port Colborne department, demonstrates a chair knot, using Ralph Cline of the English Electric department, St. Catharines, as a subject during his talk on useful knots for firemen."

St. Catharines Standard, August 22, 1949, p. 8. StCM – The Standard Collection, S1949.29.2.10

St. Catharines Standard, August 28 / 29, 1949, p. 1.
StCM – The Standard Collection, S1949.29.3.8

Sunday Fire Guts St. Paul Street Block

"Every available fireman was called out to fight St. Catharines' worst fire in eight years which raged from two o'clock until well after six yesterday afternoon. Fire started from undetermined cause in the front of MacQuillen Drug Store's basement and spread rapidly to the adjoining Singer and Desand stores. A fire-wall is credited with saving the Floyd jewellery store and its overhead apartment and dentist's office."

St. Catharines Standard, August 28 / 29, 1949, p. 8.
StCM – The Standard Collection, S1949.29.3.18

"Every available fire fighter in the city battled the conflagration that reduced the St. Paul St. building block housing three stores to a total loss. Wave after wave of dense smoke drove firemen back from buckling walls. Perched atop the aerial ladder smoke-eaters guided heavy streams of water through upstairs windows and onto the roof which eventually caved in. The fire was brought under control at six o'clock, although sodden wreckage smouldered half the night. Saved by his crash helmet from collapsing beams in the Singer store, Fireman Stu Raeburn was hustled across the road to the Fire Hall where he received first aid from Hugh McKean… of the St. John Ambulance Brigade, and from Mrs. Phil Gardner."

St. Catharines Standard, September 15, 1949, p. 1. StCM – *The Standard Collection*, S1949.29.6.1

Three Women Flee as Pre-Dawn Blaze Guts Merritton Store

"Three women sleeping in adjoining apartments escaped in night attire at 5 o'clock this morning when fire gutted the store of Mike Diacur, 86 Town Line, Merritton. Damage, partly covered by insurance, was estimated at $10,000. Fire started near [a] stock of matches in [the] grocery store and swept through apartments at [the] back and upstairs leaving only a blackened shell as shown here."

St. Catharines Standard, October 12, 1949, p. 19. StCM – *The Standard Collection,* S1949.29.7.2

St. Catharines School Children Learn Fire Prevention

"The St. Catharines Fire Department is playing an active part during the present Fire Prevention Week, in demonstrating to school children in the city the art of fire prevention, past and present. Members of the department are touring the city schools with their renovated 1908 steam pumper, the first motor vehicle pumper of 1917 and a modern fire truck. The above picture shows the 1908 steam pumper in action outside St. John's School, while pupils look on with wide-eyed interest."

St. Catharines Standard, October 12, 1949, p. 19. StCM – *The Standard Collection,* S1949.29.7.3

Leather Buckets Were Used to Fight Fires

"Deputy Chief "Red" Harris of the St. Catharines Fire Department swings to action in the above picture, as he uses leather buckets in demonstrating to the children of St. John's School, how early St. Catharines firefighters used to extinguish blazes. Fire Chief Art Burch, standing at left, spoke to children on the importance of fire prevention, and acted as master of ceremonies as pages from the St. Catharines Fire Department past were brought back to life."

St. Catharines Standard, November 13 / 14, 1949, p. 1. StCM – The Standard Collection, S1949.29.8.1

$30,000 Damage at Consolidated Truck Lines Blaze

"A general alarm fire starting from faulty wiring caused damage estimated at $30,000 to the terminal of Consolidated Truck Lines, St. Paul St. west Sunday morning. Here, black smoke pours from the building at the height of the blaze."

St. Catharines Standard, December 24, 1949, p. 12.
StCM – The Standard Collection, S1949.29.9.16

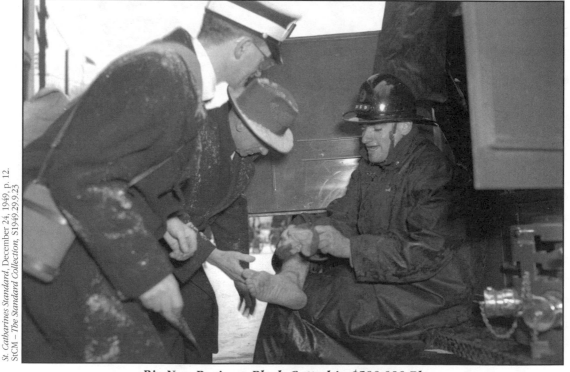

St. Catharines Standard, December 24, 1949, p. 12.
StCM – The Standard Collection, S1949.29.9.23

Big New Business Block Gutted in $500,000 Blaze

"All city firemen were summoned in a three-alarm fire which started at 9 o'clock this morning and was still burning four hours later in the fine new Burroughes Furniture Co. block on James St. Lines of hose played water on the building from all sides, in a vain attempt to save it. Here, the aerial ladder is used as a hose tower as smoke billows around it and onto uptown streets. Below, Fireman Bill Gardner receives first aid from St. John Ambulance men after he had stepped on a nail in the smoke-filled interior. His injuries were minor and he returned to duty."

*St. Catharines Standard, December 24, 1949, p. 3.
StCM – The Standard Collection, S1949.29.9.3*

$500,000 Fire Destroys Burroughes' Furniture Block

"The two-year-old stone block occupied by Burroughes Furniture Co. and the law office of Bench, Keogh, Rogers & Grass on James Street was destroyed by a general alarm fire this morning. Damage was estimated unofficially at $500,000. Fire is believed to have started about 9 o'clock in a cleaning closet on the second floor. Here, clouds of acrid yellow smoke pour from the roof of the building as all city fire-men waged a losing battle. Thousands of last-minute Christmas shoppers watched the blaze."

*St. Catharines Standard, April 22 / 23, 1949, p. 14.
The Standard Collection, S1949.28.3.1*

*St. Catharines Standard, April 22 / 23, 1949, p. 14.
StCM – The Standard Collection, S1949.28.3.3*

Life's Work Destroyed by Fatal Fire at Queenston Veteran's

"Using his gratuity from nearly five years with the tank corps of the 1st Division overseas, John Lazar, 26, bought a six-acre farm and home on Progressive Avenue a mile west of Queenston three years ago. Last night fire from an overheated furnace completely destroyed the home and all contents, and took the life of their baby, Loretta, seven months old. Loss was only partly covered by insurance. Lazar worked the farm as well as working at the Lionite Abrasives Co. at Stamford. [At left] members of the Niagara Township Volunteer Fire Department from Queenston and St. Davids under Chief Joseph Parnell, (left), wet down the fire which defied their attacks…. [Right] neighbors silently peer into the smoking foundation which still contained the body of the baby. Caleb Bennett, next-door neighbor,… was seared about the face as he tried to get in twice to save Loretta."

St. Catharines Standard, April 22 / 23, 1949, p. 1. StCM – The Standard Collection, S1949.28.3.6

Baby Dies, Queenston Veteran Loses Everything in Fire

"John Lazar, 26, veteran of nearly five years overseas, holds his daughter, Ann Marie, 3, and tried to comfort his wife after fire completely destroyed their home on Progressive Avenue, a mile west of Queenston, last evening. Their other daughter, Loretta, seven months old, sleeping in a bed downstairs, died in the fire which started from an overheated furnace. Parents were grading their driveway when they noticed the fire, but were driven back by a wall of flames when they tried to rescue the baby."

*St. Catharines Standard, April 1, 1950, p. 10.
StCM – The Standard Collection, S 1950.25.12.1*

New House Presented to St. David's Veteran After Fire

"The comfortable six-room stucco house which yesterday was presented to John Lazar, St. Davids, is shown here. It was built from public subscription on the same foundation of his former home, destroyed a year ago by fire. Lazar, wounded twice during five years service overseas, plans to hold [an] "open house" soon to let all residents of the township see the results of their generosity in coming to his aid."

St. Catharines Standard, January 13, 1950, p. 1.
StCM – The Standard Collection, S1950.30.2.1

Nearing Completion

"The St. Catharines Fire Department may move into the new central station on Geneva St. early next week, Fire Chief Art Burch stated yesterday. The interior of the $150,000 building, shown above, is virtually complete and only requires furnishing, and official inspection of the new fire alarm system. The city may well be proud of this distinctive, modern, fireproof hall which is worthy of a growing centre such as St. Catharines."

St. Catharines Standard, January 14, 1950, p. 17.
StCM – The Standard Collection, S1950.30.2.2

City Firemen Put Finishing Touches To New Fire Hall

"Mrs. Edna Lee is seen instructing Fireman Nick Batt on the operation of the new switchboard in the New Central Fire Station on Geneva St. Fireman Alphie Davis is seen at the right watching Nick plug in a practice call for the Chief. On the left is the new operator's desk with recording and transmitting equipment. On top of the desk is the tape and bell that records the box number, time and date when an alarm is turned in. On the right is a relay system for transmitting the box number to the Lake St. Station or any other station that may be opened in the future. On the left is the alarm buzzer that notifies men in the station that they are being called out. In the background is the six-circuit panel controlling the fire alarm box system."

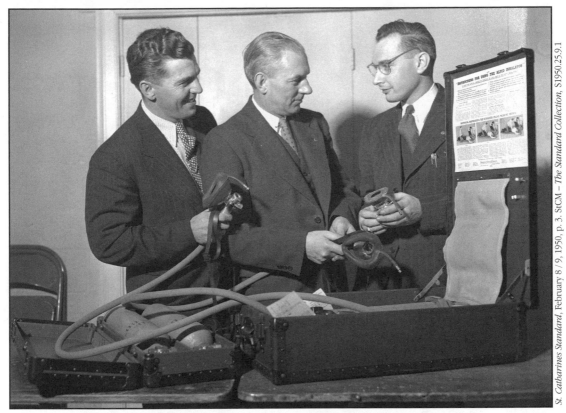

St. Catharines Standard, February 8 / 9, 1950, p. 3. StCM – *The Standard Collection*, S1950.25.9.1

Jordan Lions Give Inhalator to Fire Department

"Useful in case of water, gas, or smoke emergencies, a $500 inhalator complete with two extra supply tanks and an infants' mouthpiece, was presented to the Louth Volunteer Fire Department last night by the Jordan Lions Club at their meeting at the Jordan school. Money was raised last summer at the club's first carnival. The equipment is interchangeable with that owned by the St. Catharines and Port Dalhousie fire departments. Shown here trying out the new inhalator are, left to right, Lions Chief W. Elston Honsberger; Lion Eric Blake, fire chief; and Dr. Garnet H. Smith, chairman of the Lions Health and Welfare Committee, who made the presentation."

*St. Catharines Standard, March 23, 1950, p. 1.
StCM – The Standard Collection, S1950.30.10.2*

St. Catharines Standard, March 23, 1950, p. 1. StCM – The Standard Collection, S1950.30.10.4

Lad Already Safe; Neighbors Attempt Rescue

"Fire of undetermined origin completely destroyed the two-room winterized cottage of Mr. and Mrs. Fred Howell, Lakeside Drive, Port Weller east, shortly before noon today, with damage estimated at $3,500. Their son, Freddie, 4, temporarily left alone, was feared trapped in the inferno, but had rushed unseen from the home and was safe with neighbors. At top, firemen using water pumped from a swollen creek, wet down the ruins. Below, A. Boikoff, who lives across the road, holds Freddie. Boikoff got inside the front door after Freddie's mother screamed that the little lad was inside, but flames drove him back badly singed...."

St. Catharines Standard, March 25, 1950, p. 12. StCM – *The Standard Collection*, S1950.30.11.3.
Photo by Don Sinclair

Home of Peach Kings, Grimsby Arena Razed by $125,000 Fire

"Familiar to hockey enthusiasts throughout the Niagara Peninsula, the Grimsby Arena was completely destroyed by fire shortly after one o'clock this morning, just two hours after the winter program for the arena had closed…. some of the hundreds of spectators, attracted by a reflection visible 20 miles away, watch as fire rapidly licks up the wooden interior of the galvanized iron building…. Grimsby volunteer firemen pour water into what once was [the] front door of [the] ice palace. Fire was believed to have started after [an] explosion was heard in [the] west corner at left…. Tom Warner,… ice-maker since the arena opened in 1922, said: "it was like home to me. I was first man in when they opened it, and I was last man out last night." The big man who served with [the] R.C.A.F. security police wept openly as walls tumbled."

St. Catharines Standard, March 31 / April 1, 1950, p. 10. StCM – The Standard Collection, S1950.30.12.8

St. Catharines Standard, March 31 / April 1, 1950, p. 10. StCM – The Standard Collection, S1950.30.12.10

Firemen Put on Rescue Display in Demonstration at Fire Hall

"Culminating a week of lecture and practice in the various aspects of fire fighting and rescue work, members of the St. Catharines Fire Department yesterday put on a demonstration for members of the Fire, Light and Traffic Committee of the city council, and Ralph Leonard, from the Ontario Fire Marshal's Department. In the above photo..., George Fix is seen suspended from the aerial ladder, about to rescue 230 pound, 6 foot 2 inch Jerry Buschlen. In the photo at right, Fireman Fix is bringing Buschlen safely to earth. On the roof, Deputy Chief Jack Cropper watches."

St. Catharines Standard, May 11, 1950, p. 1. StCM – The Standard Collection, S1950.31.1.5

T.L. Yungblut's Barn Burned in Spectacular Blaze

"Fire believed started from faulty wiring completely levelled the huge rambling frame barn and three smaller buildings on the farm of T.L. Yungblut, Middle Road, Louth Township, shortly before 3 o'clock this morning. Two cows were lost, but the barn, containing 35 well-equipped stalls, was almost empty at the time. One new frame silo was destroyed, and all that remained was a pair of concrete silos. The Yungblut family sprayed their house with a garden hose and firemen from St. Catharines and Louth Township arrived quickly to wet down other buildings. The reflection was visible in Grimsby, 15 miles away."

St. Catharines Standard, June 15 / 16, 1950, p. 16. StCM – The Standard Collection, S1950.36.15.1

St. Catharines Firemen and Police Donate to Flood Fund

"From the proceeds of a softball game between the city police and firemen, the Winnipeg Flood Relief Fund became $222.72 richer yesterday. Representatives of the two bodies handed over the bundles of cash yesterday to fund treasurer, Mrs. H. J. Herrington, and in the above picture they took time out from their counting to smile at the cameraman. Pictured are, seated, Fire Chief Arthur Burch, Mrs. Herrington, and Inspector Jim Anderson. Standing at the rear, Constable John Jones and fireman Bill Bannan."

St. Catharines Standard, July 19 / 20, 1950, p. 1. StCM – The Standard Collection, S1950.31.4.10

Woman Blitz Veteran Organizes Evacuation in Hospital Fire

"Tina Rizzi, 35, cook's helper at Bellevue Lodge Convalescent Home, was heroine of the day as fire in [the] south wing yesterday afternoon forced hasty evacuation of 43 aged and helpless patients. "It was old stuff to me," said Tina, who had taken her monthly turns at fire watching in London, England, during the war. Here, she is shown comforting one of the disconsolate and bewildered men sitting on the lawn with his back to his blazing home. "There was no panic," reported Tina. "Only old Johnny there started yelling. He was worried by all the people running past his door. None of them knew there was a fire. We just told them we were taking them out to the lawn for some air." Police said they saw Tina carrying out men over her shoulder."

Fire Confined to Roof

"Damage estimated at $5,000 was done to the roof and top floor ceilings by a fire in the Bellevue Lodge Convalescent Home. Above is seen a bathroom on the second floor where the ceiling fell in. This room suffered the greatest damage in the fire which was largely confined to the roof. Fire was believed started from [an] ember landing on [the] roof after [a] patient burned cardboard in [the] basement fireplace which proprietor William Hamilton had ordered out-of-bounds."

St. Catharines Standard, July 20, 1950, p. 23. StCM – The Standard Collection, S1950.31.4.17

St. Catharines Standard, August 3 / 4, 1950, p. 1. StCM – The Standard Collection, S1950.31.5.6

$10,000 Blaze Levels Rockway Barn of Charles Bissell

"Only the silo and milk house remained after fire of undetermined origin swept the big frame barn of Charles Bissell, Pelham Road, Rockway, shortly after 10 o'clock last night. Hay and wheat from this season's crop, and some 50 loads of other hay bought recently, were lost. Here, volunteer firemen from Thorold Township, Fonthill and Fenwick, pump water into the blaze from a nearby pond dug this spring for fire protection. Charles Bissell, Rockway,... suffered painful burns to his hands last night as he moved the hot machinery from his burning barn...."

St. Catharines Standard, August 21, 1950, p. 11. StCM – *The Standard Collection*, S1950.89.5.1

St. Catharines Standard, August 21, 1950, p. 11. StCM – *The Standard Collection*, S1950.89.5.2

Firemen's School

"The Niagara District Fireman's Association completed a successful two-day school for firemen on Saturday and Sunday. Some 130 firemen from all points in the Niagara District, and some outside, attended the two-day session. The above pictures shows a group of firemen extinguishing an oil fire as one of their lessons."

"There were a total of ten first-year classes at the firemen's school, as well as four advanced classes. The art of tieing [tying] knots was one of the important instructing points, and in the above picture, a group of firemen go through the paces in their knot-tieing [tying] education."

St. Catharines Standard, September 28 / 29, 1950, p. 10. StCM – The Standard Collection, S1950.31.7.1

Loss in Hartzell Road Fire

"Fire destroyed the building formerly occupied by Jones Packing Company on Hartzell [Hartzel] Road last night. Above, the blaze is seen at its height. Owing to the head start gained by the fire, and the long distance firemen had to lay hose, little could be done to save the structure which was largely empty at the time of the outbreak."

St. Catharines Standard, October 28, 1950, p. 10. StCM – *The Standard Collection*, S1950.31.12.4

Will Furminger's Barn Destroyed

"Two ten-year-old children playing with a cigarette lighter are blamed for starting this fire that destroyed a barn and three sheds on the farm of Will J. Furminger, Scott Street, Grantham. The building was covered by insurance, but it is not expected to be sufficient to rebuild."

St. Catharines Standard, Oct. 30 / Nov. 1, 1950, p. 1. StCM – The Standard Collection, S1950.31.13.1

Firemen Test Powerful New Pumper

"More than 850 gallons of water per minute flew in a great arc across the 12-Mile Creek at Welland Vale yesterday afternoon as the city firemen tested a new pumper truck recently purchased. The new truck will eventually be equipped with a three-way radio set, becoming one of the most valuable pieces of equipment on the force."

St. Catharines Standard, December 27, 1950, p. 1. StCM – *The Standard Collection*, S1950.31.16.1

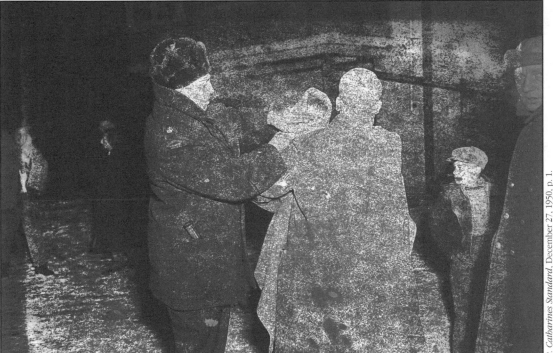

St. Catharines Standard, December 27, 1950, p. 1. StCM – *The Standard Collection*, S1950.31.16.8

Eight Flee Flames in Night Attire into Zero Weather

"Fire believed to have started from an overheated stove drove eight persons in night attire out into zero weather at 4 a.m. today when the flimsy frame duplex at 50 Kent St., beyond the city limits off Pelham Road was completely destroyed. At top, firemen fight flames after [the] house collapsed as they arrived. Below, Grantham Police Chief Norman Fach and Fire Chief Arthur Burch bundle little Sandra Tober, 14 months, in [a] blanket to be rushed to hospital in a police cruiser. Her father, Leslie Tober, was detained with painful burns to his back and shoulders after he rescued his family through a window onto the garage roof. Mrs. Tober and son Bobby, 3, were slightly burned. Downstairs, Mr. and Mrs. Frank Johnson; their son, Kenneth, 4, and Mr. Johnson's brother, Harold, escaped without injury, but lost all their belongings."

St. Catharines Standard, February 7, 1950, p. 8. StCM – *The Standard Collection*, S1950.30.4.1. Photo by Healey [W. Clifford Healey, Healey Studios, St. Catharines]

The 45 Firemen in the St. Catharines Fire Department as of February 1, 1950

"Back row, left to right, are: R. Purdy, F. Kemp, G. Hope, A. Pymont, M. Nugent, J. Wilson, G. Buschlen, F. Boyden, W. Hose, W. Betts, P. Gardner, K. Roberts, F. Peacock, W. Yaxley, J. Urquhart and F. Gill.

Second row, left to right, are: N. Batt, S. Raeburn, R. Hastings, G. Foran, A. Davis, P. Whalen, W. Bannon [Bannan], W. Brady, G. Myers, F. Shambleau, W. Gould, J. Bolan, M. Nicholls, W. Dowd, A. Smythe, J. Gill and S. Peters.

Third row, left to right, are: (front row): Fire Prevention Officer, H. Oliver; Lieut. C. Garner, Capt. J. O'Connell, Capt. B. Comfort, Deputy Chief R. Harris, Chief L.A. Burch, Deputy Chief J. Cropper, Capt. J. Murphy, Capt. R. Bracken, Lieut. T. Hunt, mechanic, H. Moore, Fire Alarm Superintendant."

St. Catharines Standard, February 2, 1950, p. 16.
StCM – The Standard Collection, S1950.30.3.2

Fire Hall Is Officially Opened

"St. Catharines' new central fire hall was officially opened at 2.30 p.m. today in special ceremonies at which his Worship Mayor Richard M. Robertson and other members of city council officiated. The Fire Department moved into their new quarters yesterday morning, and it is expected that [an] "open house" for the public will be held next week."

St. Catharines Standard, February 2 / 3, 1950, p. 1.
StCM – The Standard Collection, S1950.30.5.2

Cutting the Ribbon

"The official ceremony of opening the new Central Fire Hall on Geneva Street took place yesterday afternoon and here Mayor Richard Robertson is shown cutting the ribbon. Behind the Mayor are Ald. Gale, Barley, Wallis, Fire Chief Burch, Ald. Grammar and Moir."

SUBJECT INDEX

Decorative tympanum surmounting the doorway of the Lake Street Fire Hall, July 13, 2008.

BRIAN PHAIR

THIS BOOK HAS BEEN RESEARCHED and compiled by Brian Phair. Brian was born and raised in St. Catharines, Ontario, and attended public and high school in the city.

He believes in giving back to his community and has been a volunteer in scouting, the *Wells of Hope* in Guatemala, as a hockey team trainer, a blood donor and bone marrow registrant, and more.

In high school Brian knew that he did not want a desk job. The choice to go into firefighting was easy – it was a career where he could be active while still giving back to his community. From that point, his energy and training began to focus on this goal.

His first interaction with fire services was in 2005 on a high school co-op placement with the *Niagara Regional Children's Safety Village* in Welland, Ontario. He taught fire safety and prevention in both official languages, to school children ages 5 to 10 from across the Niagara Region. He enrolled in the *Pre-Service Firefighter Education* and *Fire Science Technology* programs at Lambton College in Sarnia, Ontario, and expects to graduate in August 2008 with a diploma in *Fire Science Technology*.

His first fires off the training grounds were in northern Saskatchewan during the dry summer of 2006 where hundreds of wildfires burned out of control. This experience galvanized his earlier thoughts about firefighting; he now knew more than ever that he wanted to fight fires. Brian's other pursuit is the military. In 2006, he enlisted in the Canadian Forces Reserves with the 4th Battalion of the Royal Canadian Regiment in London, Ontario.

This publishing project has allowed Brian to combine his interest in the past with his career goals. It is his hope that *Fire Services in Niagara* will be appreciated as a tribute to the men and women – both past and present – who put the lives of others before their own in providing for the safety of our homes and our communities.